Newton's Darkness

TWO DRAMATIC VIEWS

Newton's Darkness

TWO DRAMATIC VIEWS

Carl Djerassi
Professor of Chemistry, Stanford University
www.djerassi.com

David Pinner
Visiting Associate Professor of Drama, Colgate University

Imperial College Press

ICP

Published by

Imperial College Press
57 Shelton Street
Covent Garden
London WC2H 9HE
www.icpress.co.uk

Distributed by

World Scientific Publishing Co. Pte. Ltd.
5 Toh Tuck Link, Singapore 596224
USA office: Suite 202, 1060 Main Street, River Edge, NJ 07661
UK office: 57 Shelton Street, Covent Garden, London WC2H 9HE
www.worldscientific.com

British Library Cataloguing-in-Publication Data
A catalogue record for this book is available from the British Library.

NEWTON'S DARKNESS
Two Dramatic Views
Copyright © 2003 by Carl Djerassi and David Pinner

ISBN   1-86094-389-6
ISBN   1-86094-390-X (pbk)

Desk Editor: Tjan Kwang Wei

Printed in Singapore by World Scientific Printers (s) Pte Ltd.

Contents

*I*ntroduction: Flawed Genius

"Nature, and nature's laws lay hid in night. God said, Let Newton be! and all was light"

Alexander Pope

The name of Isaac Newton appears on virtually every survey of the public's choice for the most important persons of the second millennium. A poll published in the 12 September 1999 issue of the London *Sunday Times Magazine* ranked him first, even above Shakespeare, Leonardo da Vinci, Charles Darwin, and similar canonized figures. Among his crowning achievements was the research starting around 1670 on light and color, but he is best known for his enunciation of the laws of motion and of gravitation and their application to celestial mechanics as summarized in one of the greatest tomes in science, the *Philosophiae naturalis principia mathematica*, usually shortened to *Principia*—the first version of which was published in 1687.

Putting physics on a firm experimental and mathematical foundation—an approach known as Newtonism—earned Newton the ultimate accolade as father of modern scientific thought. A revisionist historical analysis, based in part on the discovery by the economist John Maynard Keynes of a huge trove of unpublished papers and documents, has led some scholars to consider Newton the last great mystic rather than first modern scientist. While his work in physics and mathematics set in motion the Age of Enlightenment, revisionist historians point out that neither as a person nor an intellect did he belong to it. As debunking of some of the hagiography surrounding Newton commenced in the latter part of the twentieth century, it became evident that Newton spent much more time on alchemy and mystical theology than on "science"— composing over one million words on each of these two endeavors,

—much more than all his writings on physics combined. His alchemical library was huge and his alchemical experiments, though kept secret from all but a few intimates and servants, consumed many of his waking hours for decades. The scientific genius whom Alexander Pope credited with shedding light on Nature was also a man of disturbing darkness.

The term "darkness" can be applied to much of Newton's personality. A deeply complex man, he was also morally flawed. Adjectives that have been used to describe facets of his personality are remote, lonely, secretive, introverted, melancholic, humorless, puritanical, cruel, vindictive, and, perhaps worst of all, unforgiving. Some readers have even discerned malice in the most famous apothegm attributed to Newton, "If I have seen further it is by standing on ye sholders [sic] of Giants." Often cited as a sign of his modesty, it has also been interpreted as the ultimate poisonous lacing in a disingenuously polite letter addressed to one of his bitterest scientific foes, Robert Hooke, of pronounced dwarfish stature. (It is worth noting that the origin of the aphorism long antedates Newton since it can be traced to at least John of Salisbury in the twelfth century.) But irrespective of its source, it can—in the words of Robert Merton, the master sociologist of science and author of *On the Shoulders of Giants: A Shandean Postscript*—be taken in two ways: extolling the dwarfs, who are raised high on the giants' shoulders, or the giants without whom there would be no eminence from which the little men could see far and wide.

The onset of Newton's "darkness"—so intimately related to his obsession with alchemy and his religious convictions—can be dated to Christmas Day 1642 (the year of Galileo's death) when Newton was born prematurely. Newton's father had died at the early age of thirty-six, two months before his son's birth. To make matters worse, very soon after Newton's birth, his mother married the rector of North Witham, Barnabas Smith, and left her infant son to be brought up by his grandparents. His mother only returned to the family home when Newton was a pre-pubescent eleven year-old. Newton never forgave his mother for this perceived abandonment and henceforward never trusted any woman. He loathed his stepfather; so much so, that in his *Fitzwilliam Notebook* (now in the Fitzwilliam Museum) Newton wrote, "Threatening my father and mother Smith to burne them and the house over them."

As a result, the young Newton became intensely introverted and arrogantly Puritanical. So convinced was he of his supernatural powers that he once constructed a virtual anagram of his name *(Isaacus Neutonus)* in terms of "God's holy one" *(Jeova sanctus unus)*. His own secret religious beliefs even caused turmoil during his tenure as Lucasian Professor of Mathematics at Cambridge University's Trinity College. As holder of that chair, Newton was supposed to be ordained into the Anglican Church, but as a closet Aryan regarding Trinitarianism as heresy he continually sidestepped the issue. Like the Alexandrian priest, Arius, at the Council of Nicea in 325, Newton believed that God and Christ were separate entities and that the Bishop of Alexandria, Athanasius (later beatified by the Catholic Church), was "a vile blasphemer" because Athanasius insisted that the Early Church accept the notion of Homoousion—the doctrine stating that God and Christ are of the same substance. At Athanasius's insistence, the Catholic Church— regarded by the Puritan Newton as the work of the Devil—adopted the doctrine of the Holy Trinity, as did the Anglican Church. So Newton spent several years of his life researching the history of the Church. He felt the day approaching when he might have to prove to Trinity College that the Holy Trinity (Father, Son and Holy Ghost, one and indivisible, three in one, one in three) was not only blasphemous but also mathematical nonsense. As Newton was preparing to publicly defend his heretical beliefs to the authorities, his friends at Charles II's court persuaded the king to convert the Lucasian Chair into a purely secular position and thus saved him from professional hara-kiri.

Newton's subsequent elevation to the important government rank of Master of the Mint and conferment of a knighthood by Queen Anne should have required open adherence to the Anglican Church. Yet Newton avoided confrontation with the Church throughout his adult life, and was only openly defiant on his deathbed in 1727 at age eighty-five, when he refused the last rites. Even that noncompliance did not prevent a state burial in Westminster Abbey nor the unveiling there in 1731 of a monument in just recognition of his towering contributions to science and of his services to England.

Newton's obsession with alchemy almost antedated his immersion in religious studies. In his teens, he boarded above an apothecary's shop. It was because of the apothecary, Clark, that Newton first discovered his passion for chemistry and alchemy, and displayed the beginnings of his life-long hypochondria. He could not resist concocting bizarre remedies that he used both on himself and his acquaintances. As he wrote, "My Lactellus Balsam is composed of turpentine, rose-water, beeswax, olive oil, sack and red sandal wood, and is good for consumption"—which he falsely believed he was suffering from—"and is equally efficacious when applied externally to cure green wounds and the bite of a mad dog." Apothecary Clark's stepdaughter, Catherine Storer, fell in love with him, but the teenager was so consumed by his studies and his boyhood experiments that he soon forgot his childhood sweetheart—an early sign of life-long bachelorhood.

When Newton achieved a place at Cambridge University, he continued experimenting on himself. He went almost blind looking for extended periods directly into the sun while observing colored rings and spots before his eyes. As he wrote, "I took a bodkin, and put it between my eye and the bone as near to the backside of my eye as I could; & pressing my eye with the end of it (so as to make the curvature in my eye) there appeared several white, dark and colored circles." Thus he nearly ended his scientific career before it had begun.

In the latter part of 1669, when Newton was only twenty-six, he was appointed as the second Lucasian Professor of Mathematics at Trinity College (a chair now held by Stephen Hawking). His first lecture was on optics—eventually the subject of his groundbreaking book, *Opticks*. But not a single student turned up for his lecture the following week. For the next seventeen years Newton found himself mostly alone in the class room "and for want of hearers, he read to the walls."

During his early college days at Trinity, the twenty year-old Newton, who found it very hard to mix with his fellow students, profited from his peers by becoming a moneylender. Ironically Newton's fervent interest in money did not reach its apotheosis until many years later, when he was appointed Warden (1696) and subsequently (1699) Master of the Royal Mint. (As a wayward Puritan student, Newton criticized himself in his Fitzwilliam *Notebook*, "Setting my heart on money more than God," but that did not keep Newton at age seventy-eight from speculating and losing heavily in the South Sea Bubble financial debacle.)

Then, in 1663, John Wickins—the first of two men featuring in Newton's repressed homosexual life—came to room with him. Wickins, the son of the Master of Manchester Grammar School, lived and worked with Newton for the next twenty years. They turned their rooms into an alchemist's laboratory filled with the fumes of sulfur and quicksilver. It was Wickins who observed of Newton, "endless nights he did not sleep or eat because of his desire for the Philosopher's Stone." Newton told Wickins that he attributed the prematurely gray color of his hair to so much experimentation with quicksilver that "*I took so soon the colour.*" Newton, the genius who would one day write the unrivalled *Principia*, knew only too well how he would be judged by his peers if they ever discovered that he was spending infinitely more of his time on alchemy than on science (or natural philosophy as it was then called). Hence alchemy, like religion, became a part of his obsessively secret life.

Newton felt more than friendship for Wickins, yet it is almost certain that his Puritan beliefs prevented him from indulging in physical intimacy with his friend. But Newton's vindictive nature showed itself when Wickins deserted him by marrying and becoming a clergyman. Wickins still wanted to continue their friendship, but when he asked Newton for a donation of Bibles for his flock, Newton sent him the Bibles but refused to communicate with him ever again.

The character trait most relevant to the present book is Newton's obsessively competitive nature. Frank E. Manuel wrote in 1968 in one of the great Newton biographies that "the violence, acerbity, and uncontrolled passion of Newton's attacks, albeit directed into socially approved channels, are almost always out of proportion with the warranted facts and character of the situations." This statement applies in spades to three of Newton's best-known bitter conflicts: with the physicist Robert Hooke, the astronomer royal, John Flamsteed, and a German contemporary of almost equal intellectual prowess, Gottfried Wilhelm Leibniz—the last fight eventually turning into an England vs. Continental Europe competition. It is two of these three relentless drawn-out battles that we intend to illuminate in the form of historically grounded drama. Below is presented a brief summary of the historical evidence, starting with the Newton-Hooke struggle (Chapter 2) conducted *mano a mano*, to be followed by historical background on the

Newton-Leibniz confrontation (Chapter 3), which was fought largely through surrogates.

⚜

Newton had not been exposed to any form of censure by his peers until he published his "Theory of Light and Colours" in the Royal Society's *Transactions* in 1672 at the age of thirty. A few weeks after Newton had been accepted into the Royal Society, his paper on optics was dismissed in a letter by the Society's Curator of Experiments, Robert Hooke (seven years Newton's senior), "as to Mr. Newton's hypothesis of solving the phenomenon of colours, I confess I cannot yet see any undeniable argument to convince me of the certainty thereof." Newton was apoplectic. So the battle began between two of England's greatest natural philosophers. It lasted for over thirty years, until Hooke's death in 1703.

But Hooke, whose childhood was as traumatic as Newton's, was an entirely different creature to the Puritan, over-secretive, and paranoid Cambridge recluse. Hooke's father was an unbalanced clergyman who hanged himself when Hooke was in his teens, but the youth responded to his emotional loss by utilizing his considerable talent as an artist. After receiving a small inheritance of £100, Hooke was sent off to London to be taught by the famous portrait painter, Sir Peter Lely. Not long after his arrival in London, Richard Busby, a master at Westminster, realizing that Hooke possessed a fine analytical mind, helped him secure a place at Christ Church College, Oxford, where he received an M.A. in 1663. But whereas Newton was over-sensitive and introverted, Hooke was carefree and gregarious, soon becoming involved with the Invisible College where he associated freely with the influential thinkers who eventually created the Royal Society.

In 1662, Hooke was appointed The Royal Society's Curator of Experiments, an office he filled innovatively in spite of much pressure on his time. Like Newton, he produced theories on mechanics, gravity and optics. But unlike Newton, Hooke did not concentrate on a single problem for decades until it was solved. Rather, his frenetic energy caused him to flit from one theory to another. He developed hypotheses in geology, botany, cartography and telescopes. His major work,

Micrographia, was a treatise on the microscope that also contained some original theories on the nature of light, thus crossing Newton's path. In addition, Hooke was an architect who was admired by Sir Christopher Wren. Because of his Renaissance-like genius, Hooke was known as "the Leonardo of London," but to Newton he was simply a profligate dilettante. By contrast, Hooke—a bon vivant and great frequenter of inns and coffee houses with his many friends, including Wren and Halley—despised Newton's sedentary and apparently chaste life. The latter stood in marked contrast to Hooke's numerous affairs with women, including a prolonged incestuous relationship with his young niece, Grace Hooke, who also served as his housekeeper for several years. Hooke went so far as to describe in his extensive diaries his various sexual exploits and the quality and the quantity of his orgasms, along with his bowel movements.

Newton's response to Hooke's dismissive letter of 1672 was swift: "I doubt not but that upon severer examinations, my theory on light and colours will be found as certain a truth as I have asserted it." Newton threatened to leave the Royal Society and refused to have any of his own work published for the next twenty years until the appearance of his *Principia*. This was the beginning of many feuds between Newton and Hooke, including one where Hooke accused Newton of using passages of his *Micrographia* to demonstrate Newton's opposing theory of light.

The vituperative conflict between them (to be illustrated more fully in Chapter 2 in "*Newton's Hooke*") continued over the years, finally coming to a head when Hooke confronted Newton on the subject of gravity. Hooke realized that any theory on gravity, in relation to the planets, had to be predicated on elliptical motion, but Hooke did not have the mathematical sophistication to prove it. So, in January 1680 he challenged Newton, "I doubt not but that with your excellent method, you will easily find out what that curve [the ellipse] must be, and its properties, and suggest a physical reason for this proportion [the inverse square law]." Hooke believed that Newton would fail, but Newton not only took up his challenge, it irrevocably led Newton to his proven theory of universal gravitation.

Two years after the publication of the *Principia* (1687), Newton met Nicolas Fatio du Duillier, who was soon to become the love of his life.

His frustrated relationship with Fatio, like that with Wickins, was also probably never consummated but was largely the cause of Newton's nervous breakdown in the autumn of 1693. Fatio, who came from a wealthy Swiss family, was a spoilt but charismatic youth of twenty-five when he met the forty-seven-year-old Newton at the Royal Society. Although Fatio had a great talent for mathematics and natural philosophy, he had an even greater talent for flattery and self-promotion. From Newton's letters it is obvious that the older man was besotted by the younger, with whom he was soon conducting alchemical experiments. Fatio was a spendthrift, and he encouraged Newton to provide him with money. Indeed there was very little that Newton would not do for the young protégé, which caused Newton's numerous enemies to soon dub Fatio as "Newton's ape." Despite this, Newton continually tried to persuade Fatio to move into his rooms at Trinity, but his protégé always demurred. Then Fatio became careless and began to send Newton alchemical secrets by post. The paranoid Newton, terrified that his enemies, specifically Hooke, would discover his obsession with alchemy and his unacceptable religious beliefs, broke with Fatio, whose religious beliefs were even more extreme.

Within weeks of their relationship ending, Newton's chained emotions snapped into temporary insanity. In September 1693 Newton wrote a deranged letter to Samuel Pepys; shortly thereafter, he accused his friend, the philosopher Locke, of "having endeavoured to embroil me with women." But Locke, realizing that his friend was mentally unstable, forgave Newton for his libelous accusations. Locke also knew that Newton had enjoyed serving briefly as the member of parliament for Cambridge, and that he was now desperate to acquire an eminent civil position. With the powerful assistance of another of Newton's friends, the Chancellor of the Exchequer Charles Montagu, Locke helped Newton in the early spring of 1696 to become Warden and three years later Master of the Royal Mint.

With that appointment and move from Cambridge to London, Newton left forever the world of science and alchemy and set about the recoinage of England. If Newton had failed in his new post, it could well have broken the English economy and set off a social upheaval equivalent to that of the Civil War. He applied his ruthlessness and mathematical genius to the task. Within three years he had successfully re-coined six and a half million pounds—an Olympian achievement considering that barely half that amount had been produced in the previous thirty years. To ensure the continuing success of the re-coinage, Newton personally supervised the interrogation, the trials and the executions of every apprehended counterfeiter and coin-clipper, commenting brusquely, "Criminals, like dogs, always return to their vomit."

As Master of the Mint, the academic recluse Newton turned into a figure in London society, with his niece Catherine Barton, a great beauty and erstwhile mistress of Jonathan Swift, becoming Newton's confidante and housekeeper. Unlike Hooke's incestuous relationship with his niece, Newton's affection for Catherine was purely platonic. But Newton's friend and enabler, Chancellor of the Exchequer Charles Montagu, had a passionate relationship with Catherine until Montagu's death in 1715. Hypocritically, Newton, the rigorous Puritan, turned a blind eye to their twelve-year affair even though the lovers often cohabited under Newton's own roof—because Newton never forgot that it was Montagu who had made him Warden of the Mint, the initial source of his power and growing wealth.

Newton rarely attended the meetings of the Royal Society at Gresham College because of the omnipresence of his life-long adversary, the society's secretary, Robert Hooke. Then in March 1703 the overworked and decrepit Hooke died. In November of the same year Newton, by now one of the most famous scientists in the world, was made the Society's president. Instantly he took his revenge on his dead enemy, though not openly. Instead, Newton first approved the Royal Society's acquisition of new premises in Crane Court, Fleet Street, and then arranged for the mysterious loss of the only existing portrait of Hooke, along with many instruments created by Hooke, none ever to be seen again. Now at last Newton felt free to publish his second work of genius, *Opticks*, in four volumes. It was only because of Hooke's critique of Newton's "Theory of Light and Colours" that Newton had remained silent on the subject for over thirty years. Yet even in his final years, he was often heard to mutter, "Damn Hooke, damn him."

Although Newton's wealth and prestige continued to grow apace, culminating with his being knighted by Queen Anne in 1705, even in his twilight years his appetite for vengeance remained as rabid as ever. When "Le Grand Newton"—as he was known on the Continent—was not destroying the reputations of fellow scientists, as late as three years before his death at age eighty-five he was still sending counterfeiters to the gallows without a hint of mercy. Or, bizarrely, trying to re-create the proportions of Solomon's Temple in his library because he regarded Solomon as the greatest alchemical magus of them all.

Newton's vengeful vindictiveness might be equally well illustrated by his drawn-out fights with the Astronomer Royal John Flamsteed and his protégé Stephen Gray, but the decades' long priority struggle with the German Gottfried Wilhelm Leibniz concerning the invention of the calculus differed in an important respect: it transcended personal priority claims to become one between nations that was largely conducted through surrogates rather than the principals themselves. As with Hooke, Newton pursued the vendetta beyond Leibniz's death by removing any mention of his competitor in the final revision of his *Principia*—a fate posthumously also experienced by Flamsteed.

In addition to his monumental contributions to physics summarized in his *Principia*, Newton was also an inventor of the calculus (which he first called the "method of fluxions"). Up in Parnassus or down in his grave, he would immediately interject: "An inventor? Was I not the creator of the calculus—a bedrock of modern mathematics since it first revealed the relationship between speed and area?" Why would such a genius even ask such a question? Because as we have already amply demonstrated, Newton was also a fallible human being for whom priority—and especially priority about the calculus—counted above all else.

Priority can only be assigned after a definition of the term has been agreed upon. To this day, no such unambiguous definition has been produced in science, where multiple independent discoveries occur all too frequently. In many instances, the question must be asked whether priority should be assigned to the first discoverer, to the person who published first, or to the one who first understood the nature of the discovery. In the case of the calculus, both inventors fully comprehended the nature of their invention. Furthermore, it is now clear that Newton was first in terms of conception, whereas Leibniz long predated the secretive Newton in terms of publication. But since in Newton's mind and words "second inventors have no right," resolution of that priority dispute required for him a fight to the death, like a gladiator in a Roman circus. Unlike the gladiators, Newton was a consummate master of using surrogates and continued the struggle even after Leibniz's burial in Hanover in 1716.

The calculus priority struggle—with each protagonist ultimately charging the other with piracy—has, in the words of William Broad, "been fought for the most part by the throng of little squires that surrounded the two great knights." It is through the story of some of Newton's "little squires" that Chapter 3 (in the play *Calculus*) tries to examine one of Newton's greatest ethical lapses.

The stage was set by the aforementioned Nicolas Fatio de Duillier, who became Newton's most fawning disciple. This "Ape of Newton" shot the first brutal salvo openly accusing Leibniz of plagiarism. Like Newton, Fatio never married; like Newton he indulged in alchemical experiments and religious fanaticism; but unlike his mentor he went way beyond him in that regard by openly associating with the Cevennes Prophets, who spoke in tongues and became possessed during religious ecstasies. Fatio's accusation of Leibniz was not pursued—partly because of the former's religious excesses as well as Newton's fear that their joint alchemical experiments would be exposed in the process—but in 1708, another loyal follower of Newton, John Keill (secretary of the Royal Society as well as "a war-horse, whose ardor was so intense that Newton sometimes had to pull in the reins"), formally repeated the charge of Leibniz's plagiarism—an accusation published in the *Philosophical Transactions* of the Royal Society in 1710. And when Leibniz, as a long-time foreign member of the Royal Society, demanded an official retraction, Newton in his capacity as president created a commission of eleven Fellows of the Royal Society ("a Numerous Committee of Gentlemen of Several Nations") to adjudicate the conflict. At a Royal

Society meeting on 24 April 1712, a 51-page long report—partly in Latin and replete with references to private as well as published letters and documents primarily in the possession of Newton's correspondent John Collins—was read openly (and subsequently published by the Royal Society) under the title *commercium epistolicum collinii & aliorum* ("exchange of letters from Collins and others") in which Keill's accusation was totally supported.

Such a blatantly biased procedure, though clearly to be condemned, was nevertheless to be expected, considering that Newton as president of the Royal Society had indirectly appointed the committee. But further scrutiny reveals much blacker details.

The composition of the committee that never openly signed the document did not become acknowledged for over 100 years. Not only do we now know the identity of the eleven fellows, but even more importantly their dates of appointment. The famous astronomer Edmond Halley, the physician and well-regarded literary figure John Arbuthnot, and the little-known William Burnet, Abraham Hill, John Machin, and William Jones were all appointed on 6 March 1712. Francis Robartes (Earl of Radnor) was added on 20 March, Louis Frederick Bonet (the King of Prussia's resident in London) on 27 March, and three more members, Francis Aston and the mathematicians Brook Taylor and Abraham de Moivre, on 17 April.

Why should these dates be significant? Because it is patently impossible that at least the last three members, appointed on April 17, could have had anything to do with a lengthy and complicated report *officially* presented seven days later! In point of fact, none of the eleven fellows was authorially responsible, because Newton himself had written the report! And in spite of the claim that the committee consisted of "Gentlemen of Several Nations," only two out of the eleven—Bonet and de Moivre—could be categorized as foreigners. In the case of Bonet, so little is known of him that even the Sackler Archive Resource of Fellows of the Royal Society does not contain his date and place of birth, although German and Swiss archives do shed some light on him. The question can rightfully be raised why such a diverse group of Royal Society Fellows, some of them of major distinction, should have allowed themselves to be so blatantly manipulated by Sir Isaac Newton—

ostensibly to be chosen as watchdogs and then so quickly transformed into barkless showdogs.

Calculus (in Chapter 3)—in the form of a play-within-a-play—provides some speculative insight into this scientific scandal through the personalities of John Arbuthnot, Louis Frederick Bonet, and Abraham de Moivre, with most of the biographical references firmly rooted in historical records, as is the case with *Newton's Hooke* (Chapter 2). And while the particular meeting of the playwrights Colley Cibber and Sir John Vanbrugh in *Calculus* is invented, both are historical characters whose respective plays *Love's Last Shift* and *The Relapse: Or Virtue in Danger* and their final collaboration, *The Provok'd Husband,* are part of the proud canon of British Restoration drama.

This brings us to a final comment: Why did we choose the form of two theatre plays—*Newton's Hooke* and *Calculus*—for an exploration of some of the darker aspects of Newton's highly complex personality? "What purpose is served by showing that England's greatest natural philosopher is flawed... like other mortals?" asks one of the characters in *Calculus*. "We need unsullied heroes!"

But what if the hero is sullied? At stake is an issue that is as germane today as it was 300 years ago: a scientist's ethics must not be divorced from scientific accomplishments. There is probably no other scientist of whom so many biographies and other historical analyses have been published as Isaac Newton. To this date, new biographies or books about him are published almost annually—all of them in the standard format of academic or documentary prose because of their didactic purpose to transmit or interpret historical information. But since we chose to concentrate on the *human* aspects of Newton's persona, we felt that his personality also merited illumination through the most human form of discourse, namely dialog.* Most modern plays are "played" rather than read. It is our hope that *Newton's Darkness: Two Dramatic Views* will enjoy a double role by finding a home on the stage as well as in the hands of engaged readers.

*An additional example of this approach has recently appeared in French: *la Guerre des science aura-t-elle lieu?* by Isabelle Stengers (Le Seuil, Paris, 2001), dealing with the Newton-Leibniz controversy.

*F*irst View: Two Principals

Newton's Hooke by David Pinner

**(Time: 1665–1703. Cambridge and London,
mostly in Newton and Hooke's rooms)**

CAST IN ORDER OF APPEARANCE

Sir Isaac Newton (1642–1727), England's greatest mathematician and
natural philosopher, also immersed for decades in alchemy and heretical
theology. Enunciated the laws of motion and gravitation and their
application to celestial mechanics. Made fundamental contributions to
light and color as well as inventing a form of the calculus (termed by him
"Method of Fluxions"). Author of two of the most important books in
science: the *Philosophiae naturalis principia mathematica (Principia)*
and *Opticks.* Fellow of the Royal Society in 1672, president of the Royal
Society from 1703 to 1727. In 1669 elected Lucasian Professor of
Mathematics at Cambridge University. Appointed Master of the Royal
Mint in 1699 and knighted in 1705 by Queen Anne. Notorious for
ferocious priority struggles with scientists (e.g., Flamsteed), but none
greater than the ones with Hooke and Leibniz. Buried in Westminster
Abbey, where his monument was unveiled in 1731.

John Wickins (1640–1727), Trinity fellow and Newton's assistant for
20 years. Left Newton in 1683, married, and became a clergyman in
Stoke Edith, Monmouth. *(This role can be played by the same actor as
Fatio and Montagu.)*

Catherine Bakon (1643–?), stepdaughter of the apothecary Clark and
Newton's childhood sweetheart. Later became Catherine Vincent by her

second marriage. *(This role can be played by the same actor as Catherine Barton.)*

Robert Hooke (1635–1703), dubbed as "the Leonardo of London" by his contemporaries; theorist on mechanics, gravity and optics; developed hypotheses on geology, botany, cartography, anatomy, telescopes, microscopes, and workings of engines, plus designs for flying machines. Also a painter and architect. Appointed the Royal Society's curator of experiments 1662, nominated for an M.A., Christ Church College, Oxford, in 1663, when he also became the editor of the Royal Society's *Transactions*. Published his masterwork on the microscope, *The Micrographia*, in 1665 and was appointed the Secretary of the Royal Society in 1677. Was the bitter enemy of Newton until Hooke's death in 1703.

Grace Hooke (1650–?), Hooke's niece, lover, and housekeeper. She left Hooke for a younger man in 1688. *(This role can be played by the same actor as Annie Limlet.)*

Nicolas Fatio de Duillier (1664–1753), Basel-born Swiss mathematician with Geneva family connections, had met Leibniz in Hanover, the Bernoullis in Basel, and other mathematicians before coming to London in 1687. Became known as "Newton's Ape" (between 1689 and 1693). *(This role can be played by the same actor as Wickins and Montagu.)*

Charles Montagu (1642–1715), like Newton, entered Cambridge as a fellow-commoner in 1678. He was made a Fellow of Trinity and wrote the best-selling satire *The Country Mouse and the City Mouse*. Appointed as Chancellor of the Exchequer in 1694, and in turn he appointed his life-long friend Newton as Warden of the Royal Mint in 1695. Was President of the Royal Society (1695-98). Was made first Earl of Halifax during the same ceremony at which Newton was knighted in 1705. *(This role can be played by the same actor as Wickins and Fatio.)*

Annie Limlet, a fictitious composite of Hooke's numerous mistresses-cum-housekeepers. *(This role can be played by the same actor as Grace Hooke.)*

Catherine Barton (1679–?), a great beauty of the age. Eulogized by Jonathan Swift, and was Newton's niece and housekeeper. Was the

mistress of Montagu while he was Chancellor of the Exchequer until
Montagu's death in 1715. In 1717 married John Conduitt, who became
Newton's biographer. *(This role can be played by the same actor as
Catherine Bakon.)*

Act 1, Scene 1

NEWTON's room/laboratory. Trinity College, Cambridge. September.
1665.

The curtains are drawn. The room is in semi-darkness. A chink of
sunlight, via a judiciously-placed prism, casts the colours of the
spectrum on the opposite wall. The shoddily-dressed NEWTON is making
notes on the spectrum. NEWTON, who speaks with a Lincolnshire
accent, looks much older than he is because his hair is already grey.

There is a knock on the door. NEWTON opens the curtains. The
spectrum fades. There is another knock on the door.

NEWTON: *(Irritably)* Who is it?

WICKINS: *(Off)* Wickins. You said I could call on you mid afternoon.

NEWTON: Did I?

WICKINS: *(Off)* May I come in?

NEWTON: Briefly.

(WICKINS enters. WICKINS is from Lancashire. He wears spectacles
and has a full beard.)

WICKINS: You look in your dumps, Newton.

NEWTON: What my face says is merely green froth on the pond.

WICKINS: I don't understand you.

NEWTON: Apologies, Wickins, but when I am interrupted, it's my
way. So—shall we room together?

WICKINS: You are to the point.

NEWTON: Time is the arch-enemy of man. We must devour it before it devours us.

WICKINS: Do you mind if I sit?

NEWTON: If you must.

(WICKINS sits)

WICKINS: If we *are* to room together, certain things need to be clarified.

NEWTON: Indeed.

WICKINS: I trust I'm right in thinking that you obey the College dictum, and you don't entertain bawdy females as my last room fellow, that whoremonger Thistlewaite, did.

NEWTON: On the contrary, I chastise myself for having unclean thoughts with rigorous regularity.

WICKINS: I hoped as much. But do you ever carouse to excess, Newton?

NEWTON: The only thing I ever do to excess, is think, Wickins. *(Offering a jug)* Take some water. It's good for the kidneys.

WICKINS: *(Laughing)* They said you were a strange fellow.

NEWTON: I wasn't born on Christmas Day by accident.

(WICKINS laughs as he pours himself some water.)

NEWTON: Why do you laugh?

WICKINS: Despite your efforts to be curmudgeonly, Newton, there is something vaguely likeable about you. What I'm drifting towards is... you *are* my kind of Puritan.

NEWTON: Have you ever looked into the heart of the sun, Wickins?

WICKINS: No. That way blindness lies.

NEWTON: A philosopher needs to be blinded by the truth. Eight months ago I was determined to experience the effects of highly-concentrated light upon the eye, so I fixed my gaze on a looking glass I had made, in whose reflection I captured the image of the sun, and I gazed into God's Fiery Furnace.

WICKINS: Madness!

NEWTON: Then I turned my eyes into a dark corner to observe the circles of colours. I noted how they decayed by degrees, and finally vanished. Growing bolder I repeated the ritual a second and a third time.

WICKINS: Madness upon madness.

NEWTON: Hours later I had brought my eyes to such a parlous state that I couldn't bear to look upon any bright object because I could only see the sun perpetually before me, so I shut myself in my darkened room. Then after a week I recovered my sight pretty well. But the spectrum of the sun still burns my eyes whenever I begin to meditate upon this phenomenon. Even though I lie at midnight with my curtains drawn.

WICKINS: What did you achieve by endangering your eyes?

NEWTON: I began to unravel one of Nature's truths that will provide a working foundation on which to build a revolutionary theory.

WICKINS: A new theory of Light?

NEWTON: That would be telling, Wickins. But if we come to be roommates, I may yet share my findings with you.

WICKINS: Share them with me now. As a fellow natural philosopher, I swear I'll never reveal anything you share with me.

NEWTON: Will you swear on your immortal soul?

WICKINS: Yes. I also wish to be a servant of the Truth.

NEWTON: Will you be *mine*?

WICKINS: If you prove to be a master of Natural Philosophy, I will serve you in that particular regard, yes.

(WICKINS crosses to a crucible, covered with a cloth. He pulls back the cloth.)

WICKINS: But I doubt you are a master of *this*. If you are, you're truly amongst mankind's elite.

(Hastily NEWTON covers the crucible with the cloth.)

NEWTON: Carelessness can cost a man more than his life.

WICKINS: Too late. A glimpse revealed its purpose.

NEWTON: It has other uses than those you allude to.

WICKINS: *(Shaking his head and smiling knowingly)* Now, Newton, *I* have also dabbled in…

NEWTON: Silence!

(NEWTON crosses the room, wrenches open the door, and peers down the darkened passageway. Satisfied there is no one there, NEWTON closes the door.)

WICKINS: *(Laughing)* What do you fear? This is Trinity College, not the Vatican.

NEWTON: Don't belch the stench of Rome in my chamber, Wickins. If you wish to fondle the Whore of Babylon's loins, do so with looser companions.

WICKINS: I didn't mean to offend you, but you are a nervous manikin to be frightened by spectral ears.

NEWTON: Original thoughts are scarce in every age. They must be hidden from plagiarizing scavengers.

WICKINS: *(Pointing to the covered crucible)* I'm still amazed that *you* are a practitioner of the Art.

NEWTON: We cannot be room fellows!

WICKINS: Why not?

NEWTON: You have an indiscreet mouth.

WICKINS: *(Raising his hand in protest)* I swear that I will never…

(NEWTON slams WICKINS' hand down on the Bible.)

NEWTON: Then swear on this!

WICKINS: *(Laughing in disbelief)* You *are* serious.

NEWTON: Always—where Truth is concerned. So swear, or seek another room fellow.

WICKINS: As you will.

NEWTON: Say after me; I, John Wickins, swear that I will never betray any confidences that Isaac Newton will reveal to me...

WICKINS: I, John Wickins, swear that I will never betray any confidences that Isaac Newton will reveal to me...

NEWTON: ...And if I do, may my sojourn in Hell be perpetual.

WICKINS: ...And if I do, may my sojourn in Hell be perpetual. Now I have sworn, you must tell me why you played ducks and drakes with your sight.

NEWTON: From my Optic experiments, I have revealed properties of Light that constitute the most considerable discovery... Now remember your soul is in hazard.

WICKINS: I remember. You discovered what?

NEWTON: Draw the curtains. *(WICKINS obeys)* Leave that chink for the sun to penetrate the room.

(NEWTON places the prism in the sunbeam.)

NEWTON: There! Well... what do you think?

WICKINS: What am I supposed to be looking for?

NEWTON: Can't you see how the prism separates the sunbeam's rays into a myriad of colours? And each 'colour' is the property of a specific ray that has a specific *velocity*.

WICKINS: It's miraculous! You have unweaved the rainbow. But how do you explain the nature of such things?

NEWTON: This prism may be regarded as a filter of the *different* velocities of each of the rays. The *slower* blue rays are bent more than the *swifter* red ones. Ergo: I have discovered the *varying* refrangibility of Light.

WICKINS: Your demonstration is certainly compelling.

NEWTON: I have proved that the 'pure' white light of a sunbeam is not pure at all. Rather the beam consists of a *mixture* of all the colours of the rainbow.

(NEWTON moves a piece of paper to the focal point of the prism.)

NEWTON: By removing the blue portion of the sunbeam...

(NEWTON places another piece of paper over the prism, to block out the blue light.)

NEWTON: ...The re-focused beam now appears to be reddish. It is the subtraction—or addition—of elemental colours that results in the colours that we see around us. The *apparent* colour of an object is related both to the nature of the object's surface, and to the composition of the Light striking it. So I have made the Science of Colours mathematical.

WICKINS: You have much in common with the Royal Society's Curator of Experiments.

NEWTON: What are you inferring?

(WICKINS takes a book from a shelf.)

WICKINS: You have read Robert Hooke's *Micrographia*?

NEWTON: Certainly. It's the foremost treatise on the microscope so far.

(NEWTON studies an engraving in the book.)

NEWTON: And Hooke's drawing of a magnified louse wielding a staff is remarkable.

WICKINS: That is not a staff that the magnified louse is holding in its claws, Newton. It's a human hair!

NEWTON: *(Laughing and shutting the book)* Exactly.

WICKINS: Oh I see, Newton; you were being humorous.

NEWTON: Such things happen to even me on occasions.

WICKINS: Then you do agree that Hooke has influenced your thinking on Optics because he includes original theories concerning the nature of Light in his *Micrographia*?

NEWTON: His theories are not convincing.

WICKINS: Oh come now, rightly Hooke states that Light consists of *pulses*, or very quick vibrations propagated with a finite velocity—like *waves*.

NEWTON: On the contrary, Light is the product of *corpuscular* motion. If Light *were* 'wave'-like, as Hooke conjectures, it would be deflected by any object from its straight path like sound waves. It would not form the sharp shadows that we see about us.

(NEWTON creates a hand-shadow of a rabbit on the wall.)

NEWTON: Oh I am not denying that Hooke is an intuitively clever fellow. But he doesn't have the mathematics to analyse his data. And his fanciful claims are often at odds with the facts. For instance; he wants us to believe that white Light consists of only two pure colours; red and blue alone. But wouldn't it be strange if Light was composed of waves of just *two* sizes, when there is an infinite variety of waves in the sea?

WICKINS: You do Hooke a severe injustice. He has produced hypotheses on Mechanics, Gravity, Optics, Botany, Cartography, Anatomy and Geology. He is a great authority on telescopes and the workings of engines. And even Sir Christopher Wren has praised his architectural skills. Hooke is the London Leonardo Da Vinci.

NEWTON: Hooke does not substantiate his hypotheses with irrefutable proofs. Whereas I will give birth to a Natural Philosophy, confirmed by the strongest evidence, in place of the vague conjectures that are now bandied about everywhere. And in seeking to establish scientific truth, there is no experiment, how ever dangerous to myself—*(Fastening his gaze on WICKINS)* or to others—that I will not undertake on my journey to the citadel of wisdom.

(NEWTON points to a bodkin on a table.)

NEWTON: Kindly pass me that.

WICKINS: Of course.

NEWTON: Have you ever inserted a bodkin between your eyeball and the bone, Wickins, as near to the back of your eye as you can?

WICKINS: Certainly not!

NEWTON: Then you have much to learn. See... and the pun is intended...

(NEWTON inserts the bodkin between his eye-lid and his eyeball.)

NEWTON: ...The bodkin's point is now in such a place.

WICKINS: How can you stomach to do such a thing?

NEWTON: How can I not, when the acquisition of knowledge is at stake? Now... as I press my eye with the point, so as to make the curvature in my eye, there appears...

WICKINS: Withdraw it before you blind yourself.

NEWTON: ...There appears several white, dark and coloured circles. And these circles...

WICKINS: You're mad!

NEWTON: ...Are plainest now that I continue to rub my eye with the point. Yet if I hold my eye and the bodkin still, even though I continue to press my eye with the point, the circles grow faint. Then they disappear... until I move my eye or the bodkin again.

WICKINS : You make me want to heave my dinner over my shoes.

NEWTON: *(Laughing and putting the bodkin down)* Do you still want to be my room fellow?

WICKINS: Yes. And I will assist you in your labours to explore the mysteries. As long as I don't have to go blind in the process.

NEWTON: My sentiments entirely. So let us room together.

WICKINS: *(Pointing to the crucible)* I want to serve you in the Art also. For in that, too, you may yet prove to be a Magus of the first magnitude.

NEWTON: *(Smiling)* Is that so?

WICKINS: If anyone can find the Philosopher's Stone and transform base metal into gold, it is surely one who has had the fortune to be born on Christ's Nativity.

NEWTON: Fortune had nothing to do with it. Birth is but a beginning. It is the endeavours of the rest of his life that determine the sum of a man. But I still have many sleepless hours ahead of me to scan the heavens. So I would be obliged, Wickins, if you would leave me to the advancing night.

(LIGHTS FADE)

Act 1, Scene 2

NEWTON's room/laboratory. Trinity College. Cambridge. Four years later. October. 1669.

The hangings and cushions are now scarlet in colour and rich in texture.

The door opens to reveal a HOODED WOMAN swathed in a green cloak. Behind her is WICKINS.

WICKINS: Are you sure he is expecting you?

WOMAN: *(With a Lincolnshire accent)* I'm somewhat early, that's all.

(WICKINS waits.)

WOMAN: I need to speak with Mr Newton in private.

WICKINS: Of course. But I'm surprised you weren't seen by the bursar.

WOMAN: He was distracted by his ale.

WICKINS: Sadly it is Mr Tufferton's way. *(Still standing uncertainly in the doorway)* Well…

WOMAN: I need to be alone until I speak with Mr Newton.

WICKINS: Won't you at least uncover your head?

WOMAN: My ears are cold, sir. So…

WICKINS: *(After a short pause)* As you will, miss.

(WICKINS leaves. The HOODED WOMAN crosses to one of the bookshelves and studies the titles. She sees that a notebook has slipped behind the others. She pulls the notebook out and glances through it, unaware that NEWTON is watching her from the open doorway.)

WOMAN: Oh you startled me.

(NEWTON snatches the book from her.)

NEWTON: Who the devil are you?

(The WOMAN pulls back her hood to reveal that she is in her middle twenties with luxuriant flaxen hair.)

NEWTON: Catherine!

CATHERINE: Aren't you pleased to see me, Isaac?

NEWTON: Your appearing in my rooms is provocatively dangerous.

CATHERINE: *(Smiling)* So you *are* pleased to see me.

NEWTON: That is *not* what I am saying.

CATHERINE: My legs are weary. May I at least sit?

NEWTON: All you may do is go.

CATHERINE: Are you denying that you have missed me?

NEWTON: We have no time for this.

CATHERINE: Why not?

NEWTON: If there was a modicum of sense in that pretty head of yours, Catherine, you'd realise that if you are discovered here, it will go ill with us both.

CATHERINE: I'm glad you still find me 'pretty,' Isaac.

NEWTON: What if your raging husband discovers you here?

CATHERINE: If only he *would* rage of an evening, the nights might prove to be of some interest.

NEWTON: I will not countenance lewdness in my rooms.

CATHERINE: Still the self-flagellating Puritan. My poor Isaac.

NEWTON: I am not 'your poor' anything.

(CATHERINE indicates the scarlet cushions.)

CATHERINE: So I see. The reason I came here was to convey to you my father's congratulations on your rising to the elite rank of Major Fellow at Trinity.

NEWTON: Please thank your father for his generosity, and now…

CATHERINE: *(Overriding him)* You richly deserve all the stipends that have enabled you to furnish your room so luxuriously. Though this riot of vermilion makes it more resemble a Sultan's seraglio than a philosopher's chamber.

NEWTON: So I happen to have a predilection for crimson. It doesn't mean that I have transmogrified myself into an Eastern voluptuary.

CATHERINE: You seem to have forgotten my propensity for mischief, Isaac. *(Smiling)* Mind, the profusion of crimson still gives one pause for thought.

NEWTON: Stop this salacious bantering. Where is your husband?

CATHERINE: At home in Grantham.

NEWTON: Then why are you in Cambridge?

CATHERINE: I'm visiting my ailing cousin who lives three streets north of your College.

NEWTON: I'm sorry to hear your cousin is ill but you still can't stay.

CATHERINE: Isaac, you know why I married Francis.

NEWTON: It was your choice.

CATHERINE: May I have a little of this wine? My mouth is dry.

NEWTON: It's five days old.

(CATHERINE pours some wine.)

CATHERINE: Marrying Francis was not a choice I would have made. But when you achieved your scholarship, I knew that Trinity College had claimed you for life and you would never be mine.

NEWTON: It wasn't my intention to hurt you, Catherine.

CATHERINE: Do you still make those miraculous models?

NEWTON: What purpose is this serving?

CATHERINE: Along with your poems, I treasure your ingenious windmill. In which you placed Little Juliet to turn the sails.

NEWTON: Your pet mouse is no longer relevant to our lives.

CATHERINE: True, but it pleases me that you still remember how industriously Little Juliet trotted around inside your mill.

NEWTON: Now you've finished your wine, I will escort you to your cousin's.

(CATHERINE picks up the notebook that she took from the shelf earlier.)

CATHERINE: You've not done away with your childhood notebook, I see.

(NEWTON snatches the book from her.)

NEWTON: You still pry into what does not concern you!

CATHERINE: I know I had no right to read your thoughts.

NEWTON: So why did you?

CATHERINE: Well, when I happened upon this notebook while you were living with us in Grantham, and I found it was encoded, the Devil tempted me to decipher it. But it's only now that I see your mother is the reason that we're not together.

NEWTON: My mother has had nothing to do with my dedication to Natural Philosophy. She wanted me to stay on the farm.

CATHERINE: Yet it is because of her that you have always been so uneasy with... with...

NEWTON: ...Women? Is that what you're stuttering towards?

CATHERINE: So why did you call your mother 'a whore'?

NEWTON: Stop this!

CATHERINE: And why did you write; 'my mother and that tyrant, Smith, I want to burn them and the house over them'?

(NEWTON raises his hand to strike her.)

NEWTON: You... you...!

(NEWTON lowers his hand.)

NEWTON: The Lord preserve me, Catherine, I almost struck you. *(Pause)* You want to know why I wrote those terrible things?

CATHERINE: Only if you wish to tell me.

NEWTON: As this is the last time we will ever meet, I will acquaint you with the truth.

CATHERINE: I didn't mean to slander your mother.

NEWTON: You're right to be contrite. My mother is a good though simple woman. But she's never forgiven my father for dying before I was born. Though I still don't understand what possessed her to marry that man Smith. How could she call herself a mother and set up house with *him* in another parish? She deserted me when I was only three, and left my upbringing to my grandparents. She didn't come home again until I was eleven. All those years I felt guilty. I thought she had left me because I had committed some... unthinkable deed. The Lord forgive me but I have never trusted her since. And save for the solace of my God in my desolation, only the Lord knows what would have become of me.

CATHERINE: Father says you're a genius. Are you?

NEWTON: *(Laughing)* Your father taught me much. He's a fine apothecary.

CATHERINE: So how *do* you set about solving the problems of the Universe?

NEWTON: The process isn't so grand. When I have an original thought about a problem, I commit it to a notebook, and wait until the first dawnings open little by little into the clear light of day. Then it's only a matter of time until the problem is solved. But it's always a solitary, and often a sleepless process. Now you must go, Catherine.

CATHERINE: You can at least tell me what you've discovered in your solitary, sleepless travails.

NEWTON: You wouldn't understand.

CATHERINE: *(Laughing)* Because I am a woman?

NEWTON: Because you think you want to understand, but you don't really.

CATHERINE: You once considered marrying me, Isaac. You wouldn't have done so if you had despised my intellect. So the least you can do is to reveal something of your secret world.

NEWTON: If I do; will you go, and promise never to return?

CATHERINE: *(Smiling)* I'll go, certainly.

NEWTON: You are a woman to be reckoned with.

CATHERINE: I like to think so.

NEWTON: Do you believe Descartes is right?

CATHERINE: Is this your discovery?

NEWTON: I told you that you wouldn't understand.

CATHERINE: Don't patronise me, Isaac, until you have tried me to the full.

NEWTON: Very well. Descartes asserts that all the bodies in the Universe have moved in perfect order since their creation by God, and will continue to do so until the end of time.

CATHERINE: A beautiful theory. Is it true?

NEWTON: *(Silencing her with a gesture)* According to Descartes, there is no spontaneous motion in the Universe. Space is an infinite continuum, filled with an all-pervading aether of limitless divisibility. *(With a mischievous smile)* Are you still with me?

CATHERINE: *(Returning his smile with interest)* Absolutely.

NEWTON: *(Describing Descartes' vision with his hands)* At the moment of creation, Descartes asserts that God fashioned the denser parts of the aether into the Sun, the Earth, the stars and the planets. He assembled all the pieces of the Universe together like a vast cosmic clock. Then He set His Clock to tick silently to itself forever, without any need of further refinement or repair. And since that time God has taken no further part in the destiny of His Universe.

CATHERINE: It certainly has an exquisite logic to it.

NEWTON: Yet it is flawed on two counts. Firstly; by banishing God from the daily workings of the Universe, Descartes effectively denies the existence of the Almighty Himself.

CATHERINE: That is a disturbing thought.

NEWTON: And secondly; in Descartes' Universe there are no Forces, and Motion is only relative. But I believe that Motion is *absolute*, and that Space is an empty and fixed framework in which all astral bodies move *only* under the influence of *Forces*.

CATHERINE: How can you possibly know this to be true?

NEWTON: Last year, with the aid of my reflecting telescope, I tracked the progress of comets across the firmament the whole winter long while the University and its bibulous baboons snored in their cages.

CATHERINE: *(Laughing)* You are so wonderfully intolerant at times.

NEWTON: After a long lull without seeing a single comet, my lonely nights were finally rewarded when I spied a comet, with its firedrake tail, shooting through the constellation. From this I deduced that it was impossible that all these *random* comets could move with such seeming *impunity* through Descartes' God-*pre*-arranged continuum of space.

CATHERINE: *(Moving close to him)* I forgot how passionate you could be.

NEWTON: *(Avoiding her)* I see that my dispute with Descartes bores you.

CATHERINE: On the contrary. *(In an attempt to placate his irritation)* But you have discovered much more than this.

NEWTON: You're right. Descartes also believes that the Universe is fundamentally *geometrical*. But dynamic concepts like Force and Acceleration cannot be represented geometrically. So I have devised what I call my Fluxions—or the Calculus—in order that I can address concepts like Force and Acceleration. For Fluxions are based upon the idea of considering quantities and motions—not as definite and unchanging—but as a process of originating, *fluctuating* and disappearing. *(Pause)* Imagine two quantities – for example, the *velocity* at which I walk... *(NEWTON walks across the room)*... and the *distance* I have travelled. They are, of course, related. And using my method of Fluxions, I can find the precise mathematical relationship between the velocity and the distance. Indeed, *any* alterations in natural motion can be addressed with my Fluxions. So while Descartes' description of the Universe is poetic, geometric and general—mine is complex, mathematical and precise.

CATHERINE: And Galileo? I suppose you're equally critical of him.

NEWTON: *(Laughing)* What do you know of Galileo?

CATHERINE: Father says that Galileo is a genius. *(Again moving close to him)* Like you.

NEWTON: *(Again avoiding her)* I'm glad your father puts me in such excellent company as I am in full agreement with Galileo's Principle of Inertia that states that 'A body once moved, will move forever in a straight line - unless hindered by some external cause'. But I believe that this 'external cause' is 'Force'. *(CATHERINE undulates towards him. NEWTON backs away from her and grabs his sling from his desk and desperately loads a stone into it)* So—with this principle as my stone, and mathematics as my sling, I shall use my new exact Natural Philosophy to topple the imprecise, giant hypotheses of the past. Much in the same way as David slew Goliath. Observe.

(NEWTON whirls the sling around his head.)

NEWTON: According to Galileo's Principle of Inertia, once I release the stone from the sling, the stone should continue in a straight line. Now only an *external* Force prevents the stone from shooting off in a straight line. And it's the tension in the cord that forces the stone to continue to describe this circle around my head. And I have calculated the exact form of this Centripetal Force.

CATHERINE: *(Laughing)* Father's right; you *are* a genius!

NEWTON: This is but the start. What if there were no restraining cord, but only... infinite space?

CATHERINE: Obviously the stone would fly off in a straight line through infinite space.

NEWTON: Good. So where is the cord connecting the Moon to the Earth? For when I looked out last night, the Moon hadn't flown off into space.

CATHERINE: *(Shaking her head)* It's most perplexing.

NEWTON: Like Kepler, I believe that a Power emanates from large celestial bodies by which they *attract* smaller bodies, and influence their motion. *(NEWTON is in his own world whereas CATHERINE is the smaller body that is attracted to him. She physically mirrors his theories)* Just as the stone is held by the cord, so some Force acts upon the Moon, compelling it towards the Earth. Three autumns ago I was sitting in my mother's orchard at dusk, ruminating on the question of how far the Earth's attraction might extend... when the wind stirred the laden trees around me. A large apple fell from a branch. Then I mused that the path of a falling apple is *perpendicular* to the Earth—whereas the path of the Moon around the Earth is roughly *circular*. So what is the explanation?

CATHERINE: Is it... that the Moon is so high above us... that the Earth's power to pull the Moon towards the Earth is... well, diminished in strength?

NEWTON: Now who is the genius?

CATHERINE: *(Turning away from him)* You mock me.

NEWTON: *(Without realising what he is doing, he touches her to reassure her)* On the contrary, Catherine, you understand that the Moon's *attraction* to the Earth must *decrease* because of the *distance*.

(CATHERINE tries to press her body against his but NEWTON edges behind the table.)

NEWTON: So although the Force, exerted by the Earth on the Moon, is not strong enough to compel the Moon to crash into the Earth, the Force is still strong enough to hold the Moon in its orbit.

CATHERINE: With such awesome constraints upon the Moon, it's a miracle that we have a Moon at all.

NEWTON: *(Sitting behind his desk)* Such is the wonder of God's Universe. This Force extends to the orb of the Moon. And I believe it extends beyond... even to Infinity... *(Trailing off in deep thought)* Yet there is still so much to be worked out before such a theory is complete.

CATHERINE: *(Standing behind him)* You don't always need to be alone while you theorise.

NEWTON: There is no other way.

CATHERINE: *(Touching his wrist)* I... could visit you, Isaac.

NEWTON: *(Jumping up from his chair and knocking it over)* You are a married woman, Catherine! Even if you were not, there's no space in my life for anything but endeavour.

CATHERINE: There's much more to life than work, my dear.

NEWTON: *(Replacing the chair)* Don't you understand? It is my sacred mission to elucidate the mysteries in the Almighty's Universe. No matter the cost to myself.

CATHERINE: Oh Isaac, Isaac, you should have married me.

NEWTON: But I didn't.

CATHERINE: I can still bring warmth into your desolation because other than your obsession with Natural Philosophy, *that* is what you have reduced your existence to. So, for heaven's sake, let us enjoy some happiness together.

(CATHERINE throws her arms around NEWTON.)

NEWTON: What do you think you're doing?

CATHERINE: I'm embracing the man I love.

(NEWTON struggles unsuccessfully to free himself.)

NEWTON: Release me, before we are discovered!

CATHERINE: No, it's too wondrous having my arms around you now that I *know* you're a genius who needs my help.

NEWTON: This is lunacy!

(The flustered NEWTON is still trying to prise himself free from CATHERINE's arms, so he does not hear the muffled tapping on his door, or see the breathless WICKINS appear in the doorway.)

WICKINS: My goodness me!

(NEWTON disentangles himself from CATHERINE's embrace.)

NEWTON: This is not what it seems, Wickins. Mrs Bakon was just saying her farewells and is leaving immediately for her cousin's.

WICKINS: Then I'd best escort Mrs Bakon there because Professor Barrow requires you in his study urgently.

NEWTON: At this time of night?

WICKINS: *(Amused)* This time of night seems more than appropriate for other things.

NEWTON: Keep your libidinous thoughts to yourself, Wickins. Now escort the lady home before more disaster befalls us all.

(NEWTON rushes out, then immediately returns.)

NEWTON: Did Professor Barrow say why he required me so urgently?

WICKINS: Oh it's merely the tedious matter of appointing you as the Lucasian Professor of Mathematics in his stead.

NEWTON : What?

(NEWTON whirls out, with CATHERINE in pursuit.)

CATHERINE: This is no way to say goodbye, Isaac!

(WICKINS chases after her.)

WICKINS: *(As he disappears into the passageway)* No, but it is *his* way.

(LIGHTS FADE.)

Act 1, Scene 3

HOOKE's chambers. Gresham College. London. Four years later. June. 1673.

The slightly hunchbacked figure of ROBERT HOOKE, the Curator of Experiments for the Royal Society, is swilling his mouth out with water, and spitting into a bowl.

GRACE HOOKE, his twenty-four year-old, Titian-headed Yorkshire niece, appears silently in the open doorway. She watches HOOKE performing his ablutions with affectionate amusement. She is carrying a bowl of gruel.

GRACE: I've never known a body so busy swilling his mouth out with salt water.

HOOKE: It's only to make my breath the sweeter for my favourite niece.

GRACE: Oooh you are a naughty nuncle.

HOOKE: Who has an even naughtier niece. So set your lovely buttocks upon my rampant lap.

(GRACE sits on his knee.)

GRACE: How can it be that day and night, Robert, you are so regularly rigid?

HOOKE: Your loveliness provokes me. Oh for pity's sake, quench my ardour, Grace, before I stain my breeches.

GRACE: Later, uncle, later. *(Getting off his lap)* You promised me another kind of lesson, with an instrument that is even stiffer than your poignard.

HOOKE: Such an instrument does not exist, you wilful wanton.

(GRACE brandishes his microscope.)

GRACE: Then what is this? I want a lesson with this.

HOOKE: *(Laughing)* If you insist.

GRACE: I do.

HOOKE: Open your delectable mouth as wide as you can.

GRACE: *(Pouting)* You're being a lewd uncle again just to vex me.

HOOKE: On the contrary, niece, I'm about to combine prim scientific precision of manner with extreme intimacy of content. So stop procrastinating and open your mouth.

(GRACE obeys. HOOKE puts his forefinger into her mouth.)

GRACE: *(Talking with his finger in her mouth)* And *this* is scientific?

(HOOKE withdraws his finger.)

HOOKE: Yes. Despite cleaning your teeth with salt this morning, I have still acquired this white matter on my finger. This is what grows between your front teeth and your grinders.

GRACE: Ugh!

HOOKE: Open your mouth again.

GRACE: Do I have to?

HOOKE: If you wish to have a lesson in astonishment, yes.

(HOOKE puts his other forefinger into her mouth. Then he withdraws his finger.)

GRACE: Now your *other* finger is also slimy with my spittle!

HOOKE: Hush and observe, Grace, while I mix your spittle with the white matter from your pearly teeth... like so. Now peer into the microscope.

(GRACE looks down into the microscope.)

HOOKE: What do you see?

GRACE: In the white matter, there are many... very little...living creatures who are all prettily moving. They're like your illustrations in your 'Micrographia'.

HOOKE: They are. But then—unlike Mr Newton and his elitist friends who limit themselves by only examining *non*-living matter—*(Indicating the microscope)*—I have spent much of my time on biological examination. I am fascinated by *all* the phenomena of life. One day I hope to prove that the same laws govern dead *and* living matter. Like my friend Mr Boyle, I believe that 'The eye of a fly exhibits a more curious workmanship than the body of the sun'. But as you're unwilling to relieve my loins of their pressing burden, my sweet, I shall retire to my observatory to earn my pittance in the Royal Society's service.

(GRACE gives him his gruel.)

GRACE: First you must break your fast.

HOOKE: Not until I've perused that pile of puerile papers to prevent their appearing in the Society's 'Transactions'.

GRACE: Why torment your eyes in such a fruitless cause?

HOOKE: It's far from fruitless. Because of my Herculean efforts, the Society's magazines are no longer teeming with bovine burble. Early issues were crammed with morbid accounts of werewolves, animated horsehairs and, God help us, three-headed hermaphrodites. One pissmire of a physician even published 'An account of a Foetus that continued 46 years in the Mother's womb'!

GRACE: Now don't make yourself choleric, Uncle Robert. You'll only set off those lightnings inside your brain that cause you to vomit, like you did yesterday. So calm yourself and eat your gruel like a good philosopher.

HOOKE: *(Eating his gruel)* It wasn't anger made me void my guts.

GRACE: Was it concerning all the... blood on the table?

HOOKE: I didn't intend you to clean it, my dear, but I was utterly distraught when I fled to the coffee house. I was so consumed with guilt, I couldn't sleep. *(Pushing the gruel away)* Forgive me, but I can't eat this.

(HOOKE covers his face with his hands. GRACE tries to comfort him.)

GRACE: In Heaven's name, what's tormenting you, Robert?

HOOKE: For the betterment of mankind, I used to think that *all* experiments could be justified... but now...

(Another sob racks him.)

GRACE: Share your pain with me. I beg you!

HOOKE: While you were with Mrs Comple, I continued my investigation into animal respiration. I've long believed my findings would prove beneficial for human kind. So I performed another experiment with a stray dog. Using a pair of bellows, I filled the dog's lungs with air. When I suffered its lungs to empty again, I discovered I was able to keep the animal alive—even after I had sliced open its... its...

(With his head in hands, HOOKE trails off.)

GRACE: God in Heaven.

HOOKE: Well invoked, niece. The Almighty *was* watching me as I sliced open the beast's thorax, and cut off all its ribs. Then when I opened up its belly, for some time the dog still remained alive.

GRACE: That's truly horrible.

HOOKE: *(Nodding)* But I won't torture a creature ever again. Though much human good can be achieved if I can only find a way to render the animal insensible to its mutilation.

GRACE: If you love me, Robert, you must never do such a thing again, whether the creature is insensible or not.

HOOKE: The Royal Society has asked me to repeat the experiment.

GRACE: You don't love me!

HOOKE: I *do* love you. As God is my witness; you are the one certain light in my life, Grace.

GRACE: What kind of love butchers helpless animals?

HOOKE: I know, I know, dearest. That's why I have obtained a postponement from the Society that I promise you will be permanent.

(There is a knock on the door. GRACE exits. Muffled voices are heard in the hallway. GRACE returns.)

GRACE: There's a Mr Isaac Newton who wishes to speak with you, Uncle.

HOOKE: *(Amused)* Does he now?

GRACE: Shall I inform him that other matters detain you?

HOOKE: *(Shaking his head)* No doubt he comes to beard the critic in his den. And I am vaguely curious to meet him. So bid him...

(GRACE is about to leave as the door opens to reveal NEWTON.)

NEWTON: I do believe I've graced your hallway long enough, Mr Hooke.

HOOKE: *(Rising)* Ah, Mr Newton, it's good, at last, to make your acquaintance.

NEWTON: *(Shaking HOOKE's hand)* Your humble servant, sir. *(To GRACE)* Your pardon, Miss, but it seems I know your face, yet to my certain recollection we have never met.

GRACE: Or to mine, Mr Newton.

HOOKE: Yes, it is unlikely, sir. Grace—my niece—came to Gresham College five years ago from Bradford—as my housekeeper—and as far as I know, she has never ventured up to Cambridge. *(Suddenly uncertain)* Have you, Grace?

GRACE: No, Uncle. Though given the opportunity, I would enjoy a visit. Would you care for some chocolate, Mr Newton?

NEWTON: It's your eyes, I think.

GRACE: My eyes?

NEWTON: They're strangely reminiscent of someone I once knew. My apologies for staring. It's an unfortunate consequence of countless hours scrutinising the heavens.

GRACE: I wasn't offended, Mr Newton, but complimented.

HOOKE: Grace!

GRACE: *(Smiling)* As you see, my uncle guards my honour as if it were his own.

(GRACE exits.)

HOOKE: She is a high-spirited miss.

NEWTON: So it seems.

HOOKE: *(Indicating a chair)* Your visit comes as no surprise, Mr Newton. I've been expecting you.

NEWTON: Really?

HOOKE: You've come to acknowledge your debt.

NEWTON: Debt, sir?

HOOKE: Your recent discourse was indebted to my 'Micrographia'. Particularly in regard to the periodicity of colours in thin films.

NEWTON: It is true, I was intrigued by this phenomenon, but—unlike your good self—using a 50-foot reflecting telescope I made, I have found more than a *hypothetical* explanation for the periodicity of colours. Oh I don't decry what has been done before. What Descartes did was a good step. And you have added much in several ways, Mr Hooke. If I have seen further, it is by standing on the shoulders of giants.

(HOOKE points at his own hunched back.)

HOOKE: Are you referring to *my* shoulders, sir?

NEWTON: I am not a barbarian, Mr Hooke. But I *have* seen further because I have applied meticulous measurements and quantitative analysis in place of mere imaginings.

HOOKE: Every natural philosopher needs to use his imagination.

NEWTON: A man may imagine many things that are false, but he can only understand things that are true.

HOOKE: You are insolent, sir.

NEWTON: And *you*, sir, purport to be the Royal Society's Curator of Experiments, do you not?

HOOKE: Your question is impertinently rhetorical, sir.

NEWTON: And in your prestigious post, it was incumbent upon you to analyse my paper on the 'Theory of Light and Colours'. Then you were supposed to report to the Society as to my paper's veracity, were you not?

HOOKE: That is exactly what I did.

NEWTON: Yet you confided to Lord Brounckner that you had—and I quote—'allowed not above *four* hours for the perusal of Mr Newton's paper', and even less for the penning of your answer!

HOOKE: Why should I need longer to write on Optics when I have written so extensively on in it my 'Micrographia'? As for my only taking four hours to peruse your little paper, sir; that was more than ample.

NEWTON: 'More than ample'?

HOOKE: Certainly. When I was only thirteen, I absorbed the first two books of Euclid in only *three* hours. *(Smiling)* And I'm sure that not even *you*, Mr Newton, would deem yourself to be greater than Euclid. *(Filling his pipe)* And now this trifling matter is settled, would you care for a pipe of this excellent snuff before you return to your beloved Trinity?

NEWTON: I abominate snuff! *(Vehemently)* And there is nothing 'beloved' about the Trinity because the Trinity is…!

(Aware of HOOKE's acute gaze, the apoplectic NEWTON trails off. HOOKE lights his pipe while he studies NEWTON's flushed face.)

HOOKE: I was referring to Trinity *College*, sir. Not to the Father, the Son and the Holy Ghost.

NEWTON: *(With a forced laugh)* So you were, sir, so you were. On further consideration, I will accept your offer of a pipe.

HOOKE: *(Continuing to study him)* Will you indeed?

NEWTON: Yes, but with tobacco, if you please.

(HOOKE passes NEWTON a pipe and a pouch of tobacco.)

NEWTON: Thank you.

HOOKE: I see that the concept of the Holy Trinity makes you apoplectic, Mr Newton. Now why would that be, I wonder?

NEWTON: *(Filling his pipe)* Nothing makes me apoplectic, Mr Hooke, save Envy… when it is disguised as Wisdom.

HOOKE: What exactly are you inferring, sir?

NEWTON: Surely you recall what you wrote to the Society about my discourse on Light?

HOOKE: No, but I did make a fair copy somewhere among these papers. *(Rummaging on his desk)* Ah! Here it is.

NEWTON: Would you be so kind as to read your penultimate paragraph, Mr Hooke?

HOOKE: Aloud, Mr Newton?

NEWTON: Aloud, Mr Hooke.

HOOKE: That's an eccentric request.

NEWTON: Indulge me.

HOOKE: *(Reading)* 'Mr Newton's was an excellent discourse. As to Mr Newton's *hypothesis*…'

NEWTON: *(Interrupting)* It is not a hypothesis!

HOOKE: *(Continuing with a smile)* '… Hypothesis of solving the phenomenon of Colours, I cannot see any *un*deniable argument to convince me of the certainty of his opinions. All the experiments that I have made to verify Mr Newton's thesis only prove that Light is a Pulse —or wave—propagated through a transparent medium as *I* have always asserted'. *(Amused by NEWTON's vexation)* From your growing contortions, sir, I can only assume that—once again—you are gripped by apoplexy.

NEWTON: *(Controlling himself, and returning HOOKE's smile)* On the contrary, my dear sir. I couldn't be better pleased with your response.

HOOKE: What?

NEWTON: Initially, Mr Hooke, I admit I was vexed by having to answer your misunderstandings. I blamed my own foolishness for parting with so substantial a blessing as my quiet, to run after a shadow...

HOOKE: *(Interrupting)* 'A shadow', sir?

NEWTON: *(Relentless)* ...Such as yours, sir, but upon quieter reflection of your observations, Mr Hooke, I am delighted that so acute an objector as yourself has said nothing to *dis*prove *any* part of my paper. Least of all, with your own vapid, two-colour theory of Light.

HOOKE: *(Choking on his pipe)* The Devil I haven't, sir!

NEWTON: And that is despite your fallacious claim that you have repeated certain of my experiments which you say have convinced you that my *experimentum crucis* does not prove what I have alleged.

HOOKE: Is that surprising? When the terseness of your paper, Mr Newton, led me to deduce that your *experimentum crucis* was more likely to be fictitious than crucial.

NEWTON: What perfidy! By your own admission, you took only days to test my complex data, whereas several weeks are needed to do so in earnest. So I am equally assured that upon a severer examination of my *experimentum crucis,* you will find it to be as certain a truth as I have stated. With that, Mr Hooke, I bid you good day.

(HOOKE bars NEWTON's way.)

HOOKE: Not so fast, hot sir, not so fast!

NEWTON: *(Smiling)* It seems to me that you're the one who is now convulsed with apoplexy.

HOOKE: Jibe all you will. But isn't it your inviolable creed, Mr Newton, never to mix hypothesis with scientific fact?

NEWTON: Absolutely. What's more, you and your friends, Sir Christopher Wren and Mr Halley, would do well to employ deductive reasoning, and work *alone*—as I do—instead of forever babbling your *un*proven theories to each other in your coffee house on the Strand.

HOOKE: The last thing that Sir Christopher, Mr Halley and I do is 'babble', sir! We follow in the footsteps of Sir Francis Bacon who declared that the advancement of scientific knowledge demands the *joint* labours of *many* men to apply the experimental method to the *entire* range of Nature's wonders.

NEWTON: Yes, but you three do nothing but publicly trumpet your hypotheses like flatulent gluttons who have just consumed a wagonload of Jerusalem artichokes. *Hypotheses non fingo.* Ergo: whatever is *not* deduced from the phenomena is to be called hypotheses.

HOOKE: Yet *you* state, Mr Newton, and I quote... *(HOOKE reads from NEWTON's paper)* ...'I shall not mingle conjectures with certainties by *hypothetically* speculating on what Light is'.

NEWTON: *(Smiling)* I always practice what I preach, sir.

HOOKE: But in the very same discourse, you also claim that 'It can no longer be disputed whether Light be a Body'. Even though it has *not* yet been definitively *proven* whether Light is wave-like—as I believe; or corpuscular—as you *hypothesise*.

(HOOKE chuckles at the glowering NEWTON.)

HOOKE: Oh my guardian angel! I do believe I have reduced the great Cambridge Magus to a welcome state of dumbness. As for your 'giant' reflecting telescope that you boasted of earlier; I utterly reject its necessity.

NEWTON: You reject its necessity!

HOOKE: Moreover, I affirm that nine years ago, I constructed a little tube an inch long, to put in my fob, which performs more than any 50-foot telescope made after the common manner.

NEWTON: Mine was not made 'after the common manner', sir! I am an expert grinder of lenses. And if your 'little fob tube' is so miraculous, why have you failed to produce it at the Royal Society where you could have blinded us all with your miniaturist genius?

HOOKE: Unfortunately when I was appointed to supervise the rebuilding of London after the Great Fire, I mislaid my telescope in the ensuing turmoil. I lacked the time to prosecute my invention, and I was unwilling that the City glass-grinders should know anything of my secret.

NEWTON: *(Laughing)* So your dwarfish tube is tragically lost to posterity. Along, I surmise, with many of your other 'secret' inventions. Like your 'thirty different means of flight' that also you have refused to go into details about lest your fantastic techniques should be 'stolen'.

HOOKE: You dare to mock my accomplishments, Mr Newton.

NEWTON: I'm merely helping you to re-discover the truth, Mr Hooke. Furthermore, before I've finished with you, you will praise the brilliance of my telescope as fulsomely as Mr Huygens of Amsterdam has done. And he commands far greater respect among natural philosophers than you, sir.

HOOKE: Bilge! Everyone knows that *I* invented the spring-balance watch, not that posturing Dutch mountebank!

NEWTON: Be that as it may. Given time, you will have no choice but to acknowledge the originality of my *proven* 'Theory of Light and Colours'.

HOOKE: I will see you in the bowels of Hell first, sir.

NEWTON: And *I* will see that you sojourn there—alone, sir.

(NEWTON leaves, and returns immediately.)

NEWTON: In the interim I will inform the Royal Society that I no longer wish to be one of its Fellows. And I will never publish another discourse on Optics—or anything else!—until I have ensured that my every word is irrefutable.

(NEWTON leaves, slamming the door behind him. HOOKE begins to laugh as... the LIGHT FADES.)

Act 1, Scene 4

NEWTON's room/laboratory. Trinity College, Cambridge. Two years later. April. 1675.

Smoke billows from a crucible, enveloping WICKINS who is coughing and sneezing.

WICKINS: This is insupportable, Isaac!

(NEWTON appears with a ladle.)

NEWTON: What is?

(WICKINS watches NEWTON pour the contents of his ladle into the crucible.)

WICKINS: My nose is on fire. These filthy fumes have alchemized my lungs.

NEWTON: The Art is not to be shouted abroad. Eavesdroppers will hear us.

WICKINS: *(Sneezing)* We have not slept for three nights. *(Indicating the crucible's fumes)* My head's bonging like a belfry.

(NEWTON picks up a bottle filled with black liquid, and pours most of it into a glass.)

NEWTON: Drink a quarter of a pint of my Lactellus Balsam. It'll douse you in sweat and cleanse you of all your ills.

(WICKINS backs away from NEWTON.)

WICKINS: No! Your potion is in the latter stages of putrefaction, Isaac. What horrors does it contain?

NEWTON: *(Amused)* The health-giving Newton Formula: rose water, beeswax, olive oil burnt sack, red sandalwood, St John's wort, and a liberal sufficiency of turpentine. It does wonders for the guts, bladder, head, and testes. My brew is equally efficacious when applied to green wounds and the bite of a mad dog. Consume my elixir of life and you will be a new man.

WICKINS: Or a dead one!

NEWTON: Do you decline the fruits of my mind, John?

WICKINS: Forgive me, Isaac, but these vile brews you concoct to indulge your hypochondria will be the death of us both.

NEWTON: The last thing I am is of a hypochondriacal disposition. Still, as my grandmother used to intone: 'Waste not, want not'.

(NEWTON swallows the potion.)

NEWTON: But we have more serious matters to hand. We must produce the Star Regulus of Antimony.

WICKINS: Again?

NEWTON: Again. Oh I'm aware that the Regulus is not the Philosopher's Stone, and will never turn base metal into gold. But it is a gateway to the truth.

WICKINS: Which particular 'truth' do you have in mind?

(NEWTON stirs the crucible with a ladle.)

NEWTON: When iron, copper, tin, lead, and antimony are put into the crucible, and after fusion have stood for some time before they are poured off, there is that moment of revelation when the Regulus blossoms into a Star. Your drawing illustrates it well. Where is it?

(WICKINS rummages through some boxes.)

WICKINS: Somewhere here. Though I wish I had but half of Hooke's ability as a draughtsman.

NEWTON: Don't speak of him!

WICKINS: Isaac, lately Hooke has been most fulsome in his praise of you.

NEWTON: Pah! Beneath his honeyed words there are more barbed stings than in my mother's hives.

(NEWTON picks up a letter and waves it under WICKINS' nose.)

NEWTON: Especially in his latest letter with its insidious challenge.

WICKINS: Still, I'm glad that you no longer denigrate Hooke. At least not with your public utterances.

NEWTON: It's taken great forbearance on my part, believe me. I would have published my new theories on Optics long ago. But that's what the poisonous spider hungers for as he spins in his malignancy. So I have sworn never to publish anything until my every proof is irrefutable. *(Drinking from a tankard)* This beer is sour.

WICKINS: But you must still publish the formulae on your Fluxions before it's too late.

NEWTON: Too late?

WICKINS: Absolutely. For the last four years Collins has pleaded with you to publish your Fluxions—or the Calculus, as Leibniz now calls it.

NEWTON: *(With a barking laugh)* I must publish because of Leibniz?

WICKINS: Yes. Leibniz is about to publish his own Calculus. Oh I know his notation is different to yours, but your techniques are fundamentally the same.

NEWTON: So where is the problem?

WICKINS: Isaac, if you don't publish *before* Leibniz, he will take credit for your invention. *(NEWTON emits another barking laugh)* What's so droll about that?

NEWTON: I've already sent Leibniz two extensive letters, with an encrypted version of my Fluxions in form of a code.

WICKINS: How will that prevent Leibniz from stealing your glory?

NEWTON: If his mathematical acumen is so prodigious, Leibniz will easily break my code. Then he'll realise that *I* invented the Fluxions over a decade ago. So Leibniz will have no recourse but to acknowledge his debt to me. But enough of this. Where is your drawing of the Star Regulus?

WICKINS: I have misgivings about Leibniz acquiescing so easily.

NEWTON: You have a disturbingly suspicious nature, John.

WICKINS: *(Laughing)* Coming from *you*, that's truly wondrous.

NEWTON: The drawing, if you please!

WICKINS: *(Rummaging in another trunk)* Here. It's crumpled.

(NEWTON spreads the drawing on an easel.)

NEWTON: No matter. It is as I remember. Yes...*(Gazing at the drawing)* ...the Star Regulus of Antimony is like a giant star in God's firmament. *(A giant Star of Antimony appears behind NEWTON on the darkening cyclorama)* These shard-like crystals may be imagined as lines of light, radiating from the centre of a star.

WICKINS: Why do you find that so compelling?

NEWTON: The Star Regulus of Antimony is another image that will assist me on the path to defining a Law of Universal Gravitation.

(NEWTON uses the drawing to illustrate his thoughts.)

NEWTON: Instead of imagining these lines of light as if they are radiating *from* the centre of the star, I want you to visualise them as lines of Gravitational Force, pointing *inwards* towards the star's centre. Thus they mirror the attraction and force of Gravity. *(WICKINS laughs)* Why do you laugh?

WICKINS: Your mind is never at rest.

NEWTON: How can it rest? When there is far more in the Universe than man will ever understand. I am the last of the Magi. I aim beyond the reach of human art and industry. I've made it my business to become the world's expert on Alchemical writing because I am convinced that the transmutation of metals, and indeed of all substances, must be probable.

WICKINS: Why hazard your health on Alchemy when there is so much Natural Philosophy as yet unfathomed?

NEWTON: Alchemy and Natural Philosophy are bedfellows. The rigorous discipline and patience that are required to reveal the secrets of the crucible, are also imperative for the exploration of Science. Alchemy and Science feed into one another like streams, co-mingling into the sea of knowledge. But if you've grown doubtful as to the fruits of Alchemy, John, why do you help me?

WICKINS: I know no other life. I wish to Heaven I did.

NEWTON: *(Disconcerted)* Are you thinking of leaving me, then?

(WICKINS turns away.)

NEWTON: After all these years? *(Remembering)* Yes, yes... lately you often journey from Cambridge. As to where, I don't know. It's not in my nature to pry.

WICKINS: Would my leaving you truly upset you?

NEWTON: Yes, I would find your leaving unthinkable. Why didn't I see that such a thing was possible?

WICKINS: Sometimes you don't see me at all, Isaac. Or anyone else.

NEWTON: I live in a world of non-living matter, John. Much of the time people do not impinge upon me—save, momentarily, as random comets flaring through the night sky.

WICKINS: *(Laughing)* With firedrake tails, I have no doubt.

NEWTON: *(Trying to remember)* Yes, once I described a comet like that to... *(Trailing off)* But now the face of the listener is... veiled. Whereas your face is as luminous as the moon.

WICKINS: *(Laughing)* A compliment indeed. Oh Isaac, Isaac.

(Still laughing affectionately, WICKINS embraces NEWTON, who returns his embrace. The MEN clasp one another with great intensity.)

NEWTON: My dear, my dear...

(Abruptly NEWTON breaks away.)

WICKINS: What's wrong, Isaac?

NEWTON: Impure thoughts are the Devil's playground!

WICKINS: Impure...?

NEWTON: We who search for the Philosopher's Stone, John, must adhere to a life of chastity.

WICKINS: Isaac...

NEWTON: *(Overriding him)* We must focus our entire beings upon solving the riddle of the Universe. Its vast secrets can only be read by applying *pure* thought to the evidence that God has hidden like sapphires about the world. For the Universe...

WICKINS: *(Interrupting)* Isaac, for God's sake, face yourself!

NEWTON: ...The Universe is a cryptogram set by the Almighty!

WICKINS: Stop it, Isaac, stop this and FACE YOURSELF!

NEWTON: And I must solve the cryptogram.

(WICKINS wrenches open the door and is about to leave. NEWTON bars his way.)

NEWTON: Don't leave me!

WICKINS: Why shouldn't I leave you?

NEWTON: It's not as you think.

(Pause.)

WICKINS: Then what?

(NEWTON slumps into a chair.)

NEWTON: I may not be Lucasian Professor of Mathematics at Trinity College for very much longer.

(WICKINS closes the door and crosses to NEWTON.)

WICKINS: But you were certain that His Majesty would view your application for a special dispensation with beneficence.

NEWTON: Isaac Barrow is His Majesty's Chaplain, and he said that if the news were excellent I would learn within a short while.

WICKINS: You went to St James Palace only three weeks ago. By royal standards, that is a very short while indeed. *(Putting his hand on NEWTON's shoulder)* You must be patient, Isaac.

(NEWTON breaks away.)

NEWTON: If the King refuses to grant me a special dispensation, I will resign my Chair rather than take holy orders as 'Trinity' College demands.

WICKINS: How will you make a living without your College income?

NEWTON: Heaven only knows. But for too long now, John, I've been on the slippery slope. I have never told you, nor any man, but when I obtained my Bachelor of Arts it was through gritted teeth that I pledged to uphold the Thirty-Nine Articles of the Anglican Church, because they are the legal requirements of this damnable College.

WICKINS: Gently, Isaac, gently. Even closed doors have ears.

NEWTON: I don't care who hears me! Why should I? Then when I was appointed Lucasian Professor of Mathematics—in order to fulfil Trinity's requirements!—I promised I would take Holy Orders in the near future. To avoid committing such a blasphemy, these past six years I've procrastinated to obtain more time. But now every tongue is wagging, and time has run out.

WICKINS: I don't understand. You already live like an abstinent monk. What will you lose by being ordained?

NEWTON: The sanctity of my immortal soul.

WICKINS: How can taking holy orders in the Anglican Church threaten your immortal soul?

NEWTON: Why do you think I haven't slept for weeks?

WICKINS: You've been a slave to the crucible, reading texts of...

NEWTON: *(Cutting him)* ...*Biblical* scholarship. In case I'm challenged on my religious beliefs.

(WICKINS points to the crucible.)

WICKINS: I trust to Heaven that the Art has not turned you heretic!

NEWTON: It is the *Trinity* that is heresy.

WICKINS: In God's name...

NEWTON: Exactly. God said 'Thou shalt have *no other* gods before me'. So there is only *one* God. And not *three*, as the idol-worshipping *Trini*tarians would have us believe, with their 'God the Father, God the Son and God the Holy Ghost'!

WICKINS: But the Anglican Church places the Trinity at the centre...

NEWTON: The Trinity is the work of that idolater, Athanasius, Patriarch of Alexandria, who was part of the perfidious Council of Nicaea in 325 which condemned the great Arius.

WICKINS: Arius?

NEWTON: Yes! It was only Arius who upheld the truth that Christ is *not* God.

WICKINS: On what authority could this... Arius have made such a blasphemous claim?

NEWTON: On Saint Paul's authority, who wrote, 'There is One God, and one Mediator between God and Man—the *Man* Jesus Christ'. So we should only pray to God the Father because Christ was not divine.

WICKINS: That's heresy!

NEWTON: No, it's the Trinity that is heretical. Yet not withstanding, the corrupt Athanasius and his fellow deviants exiled Arius as a heretic. Then they cynically enforced the Trinitarian unification of the Church. Even though Trinitarianism defies fundamental logic. The principle that *three* equals *one* is no more applicable to rationalist theology than it is to the arena of mathematics. So it's simply unthinkable for me to be ordained.

WICKINS: Why do you confide this to me, Isaac? Aren't you concerned at what I could do with all that I've heard?

NEWTON: Do you wish to play Judas, John?

(WICKINS laughs.)

NEWTON: You laugh at the strangest things.

WICKINS: It wasn't *I* who said, 'I was not born on Christmas Day by accident'.

NEWTON: *(Appalled)* Are you implying that *I* think that I am another... well, another Messiah?

WICKINS: I'm implying nothing, Isaac. But I fear for you if this ever comes to light.

NEWTON: If the King refuses to grant me a special dispensation for my Lucasian Chair, I'll have no recourse but to challenge this heretical hoax.

WICKINS: Then the Lord help us both, for they will also indict me as your friend and confidant.

(There is a knock on the outer door.)

WICKINS: *(Jumping)* In God's name...!

NEWTON: Don't blaspheme. Now calm yourself and discover who it is.

(WICKINS goes off. Muffled voices are heard. WICKINS returns with a letter.)

WICKINS: It has the Royal seal, Isaac.

NEWTON: I will wrestle to establish the truth until my last breath.

(NEWTON breaks the seal, and peruses the contents.)

WICKINS: Well?

NEWTON: *(Barking laugh)* The *one* and *only* God is all-powerful and *alone* in His Heavens because Christ is only a *man*!

WICKINS: *(Delighted)* The King has granted you a special dispensation.

NEWTON: Yes, and Trinity College is now Trinity College in *name* only—until the crack of doom.

WICKINS: Let me see.

(NEWTON passes WICKINS the letter.)

NEWTON: From this day forward the Lucasian Chair is a purely secular position. Now His Majesty has ensured that my secret is safe, I am free to complete my life's work. *(Chuckling)* How my enemies would bare their fangs at this—if they but knew! Forgive me, Lord, but this makes me infinitely merry.

(WICKINS is convulsed by a coughing fit. NEWTON slaps WICKINS' back.)

NEWTON: Whatever ails you, man?

WICKINS: The damnable fumes from your crucible will consign me to a graveyard.

(NEWTON slaps his back again.)

WICKINS: No, leave me be. I need April air in my lungs and time to recover. Alone!

NEWTON: If you're certain...

WICKINS: I *am* certain. And you should read the King's missive again for fear you've misread it.

NEWTON: What?!

(WICKINS stumbles out of the room. NEWTON re-reads the letter, then brays triumphantly. The smoke from the crucible thickens. NEWTON's laughter subsides as he peers into the dense smoke. Reassured that his eyes are playing tricks with him, NEWTON subsides into a chair and closes his eyes.)

NEWTON: Sleep… that is what I so desperately need…

(NEWTON falls asleep. A COWLED FIGURE materialises in the crucible's fumes. NEWTON wakes up.)

NEWTON: John? Is that you?

(The FIGURE emits a whispery laugh.)

NEWTON: Stop playing phantoms, John. I don't believe in them.

FIGURE: But you do believe in the voices that you hear in your dreams.

NEWTON: Is this… then… a dream?

FIGURE: Your triumph is short lived.

NEWTON: What do you mean?

FIGURE: I'm sure you recall the solution you sent me, concerning the path that an object would take, when it was dropped from a high tower—as it fell to the Earth.

NEWTON: Who… are you?

HOOKE: Your answer was wrong.

(The FIGURE pushes his hood back to reveal…)

NEWTON: Hooke!

HOOKE: Yes, I am the roiling frenzy inside your skull.

NEWTON: There were no flaws in the theory I sent you!

HOOKE: You still affirm that a falling object would land slightly to the *east* of the tower?

NEWTON: Precisely.

HOOKE: But this is *only* true if the tower stood 'precisely' on the *Equator*. Whereas it is certain that if the tower was in London—or even in your beloved Cambridge—the object would fall slightly to the *south* of the tower, rather than to the east.

NEWTON: God damn you, sir!

HOOKE: I suppose that's the nearest thing to a compliment I shall ever receive from you. Mind, when I read your letter out to the Royal Society, they were even more fulsome in their praise of me.

NEWTON: You promised you would not break our pact to only discourse in *private* for the furtherance of science.

HOOKE: You can never abide to be wrong, can you?

NEWTON: If you were anything but a nightmare figment in my head, I'd swat you like an excremental horsefly!

HOOKE: Demonic pride is a sickness with you. As it was two years ago, when you resigned from the Royal Society in another of your choleric fits. Not that your resignation had any substance because three days later your friends had no trouble in persuading you to return.

NEWTON: I have never set foot within the portals of the Society since that day. And I never will—as long as *you* walk the earth.

HOOKE: Well, we must not have you idle in your self-enforced confinement. So will you now respond to the Challenge in my last letter? The Answer to which will be of the greatest concern to mankind for the invention of the Longitude will be one of its necessary consequences.

NEWTON: *(Laughing derisively)* Will it indeed?

(Behind the ghostly figure of HOOKE the stars appear on the cyclo-rama.)

HOOKE: Yes. Kepler's First Law states that the paths of astral bodies around the Sun are Ellipses. But what of the Gravitational Force between them? I challenge you to confirm *my* theory that the Force acting between, say, the Earth and the Sun depends on the Inverse Square of their separation. So you must compute the path, followed by the Earth around the Sun, under such a Force. If it is an Ellipse, as I predict, it will prove the Universal nature of Gravity.

NEWTON: Ah. *(Pause)* You have glimpsed the grandiose end then.

HOOKE: I could, of course, work out the curve myself, but I'm so overburdened in my new post as Secretary of the Royal Society that I lack the time to busy myself with the drudgery of the calculations.

NEWTON: *(Barking laugh)* Now isn't this very fine, sir? We mathematicians find out and do all the business, but we must content ourselves with being regarded as 'drudges', yet you 'overburdened hypothesizers' still have the effrontery to say that *you* are the inventors.

HOOKE: *(Amused)* So you *do* accept my challenge? I am mightily heartened.

NEWTON: Get out of my mind, you insidious vampire!

HOOKE: You will accept my challenge. Even at the expense of your health. Your compulsion to prove the truth is the only thing you have in life. It is your one fabulous gift. And life-long curse.

(HOOKE's image begins to recede.)

NEWTON: There must be an end to the need to know everything.

HOOKE: There is—but not until the yew tree's roots embrace your calcified skull.

NEWTON: I'll send you packing to Hell myself!

(NEWTON pursues HOOKE into the smoke. A moment later, with a bewildered cry, WICKINS lunges out of the fumes. WICKINS is wrestling with NEWTON. NEWTON has his hands around WICKINS' throat.)

WICKINS: Merciful Christ, you're choking me, Isaac!

(Appalled, NEWTON releases WICKINS.)

NEWTON: The Lord forgive me.

WICKINS: Whatever possessed you?

NEWTON: For a moment I thought you were… *(Trailing off)* But you wouldn't believe me. All I know is that… lately… faces tend to blur in my mind like shadows at the dark of the moon.

WICKINS: So my face is no longer 'as luminous as the moon itself'?

NEWTON: Don't mock me, John. Do you think it delights me that only equations have any tangible reality for me?

WICKINS: Yes.

NEWTON: Believe me, it's not always a blessing to have this obsession to mathematically describe the celestial dancing of the stars. Especially when it blinds me to the existence of everyone around me. Even in my deepest dreams I'm haunted by the millions on millions of pinpricks of light in the polar darkness. Then, exhausted, I wake from trying to equate how all those pinpricks of light are inter-related. Yet what else can I do, when this sacred compulsion defines my very existence?

WICKINS: As with your genius, Isaac, you're immoderate in everything you do.

NEWTON: Do I... repel you, then?

WICKINS: No. But my time with you is coming to a close.

NEWTON: Now my Lucasian Chair is safe, I assumed you would stay.

WICKINS: That's the sadness of being with you, Isaac. You assume everything will always remain as *you* want it to be—until it's too late.

NEWTON: I can change! I'm certain I can change.

WICKINS: No, no. You're not as other men, Isaac. The infinity of your mind knows no rest. And it allows none for others.

NEWTON: You know the depth of what I feel for you, John. I... I love...

WICKINS: *(Silencing him)* Once said, it cannot be unsaid.

NEWTON: I don't care!

WICKINS: Heed your religion.

NEWTON: Am I not allowed to love as others do?

WICKINS: You have no time for love. The Universe is the Almighty's Cryptogram, remember. And you were born on Christmas Day to solve it for mankind.

NEWTON: You're leaving me with nothing, John… nothing but the loneliness of the Universe.

WICKINS: It was more than enough for God's Son. It has to be more than enough for you.

(WICKINS leaves NEWTON as… the LIGHTS FADE.)

END OF ACT 1

Act 2, Scene 1

COLLINS' book shop. London. Thirteen years later. October. 1688.

HOOKE (who looks frailer and more hunched) is sitting on a stool, riffling through a book. His walking stick is propped against a table that also has a large globe on it. GRACE, his niece, who is with HOOKE, is bored.

HOOKE: Hm! His genius is unquestioned, but he's still a plagiarising pimple on the face of Creation.

GRACE: *(Yawning)* Who is?

HOOKE: The niggardly Newton! This is the third volume of his *Principia Mathematica*. And as with the other two volumes, there's no acknowledgement of his debt to *me*.

GRACE: You promised you'd buy me some ribbons for my hair.

HOOKE: He doesn't even mention my *name*, God damn him!

GRACE: Can't you buy the book and read it at your leisure?

HOOKE: You talk of 'leisure'? When Newton's every stolen word stretches me on the rack of torment. *(GRACE yawns)* Stifle your yawns, girl, 'till I'm done or, by Heaven, I'll...!

GRACE: *(Interrupting)* Don't threaten me, *old* man—for that's what your bitterness has reduced you to.

(HOOKE slams the book down on the table. Then he grabs his stick and hobbles towards GRACE, waving it.)

HOOKE: Don't tempt me, slut!

(A MAN—who is played by the same actor who portrayed WICKINS— appears from behind the shelves. But unlike WICKINS the MAN is beardless, and he is dressed like a Continental dandy. The MAN snatches the stick from HOOKE.)

MAN: *(In a Parisian accent)*   To threaten such a beautiful lady, monsieur, is a dastardly affront.

HOOKE: This doesn't concern you, sirrah.

(The MAN ignores HOOKE and kisses GRACE's hand.)

MAN: Allow me to present myself, mademoiselle; I am Nicholas Fatio de Duillier, at your service. And it is my desire that you call me 'Nicholas'.

GRACE: *(Flirting)* Then, Nicholas, you must call me 'Grace'.

HOOKE: How dare you behave like this in public, niece? It isn't maidenly.

GRACE: Is what I do with you in private, uncle, 'maidenly'?

HOOKE: I'll have no more of your whoring ways, d'you hear me?

FATIO: Shall I beat this old lecher, Grace?

GRACE: He isn't worth you bestirring yourself, Nicholas. *(To HOOKE)* But as you have brought matters to a head, you should know that I find you testy and sour-breathed, Mr Hooke. And in Spittlegate I've caught myself much younger fry upon *my* hook—who even now awaits my coming. *(To FATIO)* My thanks again—Nicholas—for your gallantry.

(GRACE leaves.)

HOOKE: *(Calling after her)* You'll be back, you strumpet! You know that none can keep you as well as I do.

FATIO: You do not deserve her, monsieur.

HOOKE: Only God can judge what any man deserves, sir.

FATIO: Hm! You are of little worth. I will leave you to choke on your dyspepsia.

(FATIO moves behind the shelves. HOOKE subsides on a stool and thumbs through the Principia. *He is not aware that NEWTON is watching him with amusement.)*

NEWTON: The treasure trove you're clutching only exists because Mr Halley came to Cambridge, and commissioned me to mathematically elucidate the Universe.

HOOKE: So the plagiariser has come in person, has he, to view how his tawdry wares are selling?

NEWTON: I've never plagiarised in my life, sir. I leave that to those *without* mathematics.

HOOKE: Cavil all you will, sir. If you had not taken up the Challenge in my letter, you would never have written your masterwork, the *Principia*.

NEWTON: Oh come now, there is a universe of difference between a truth intuitively glimpsed, Mr Hooke, and a truth mathematically demonstrated.

(NEWTON takes the Principia *from HOOKE.)*

NEWTON: Besides, everyone knows that I have gained my recent honours by my own Olympian endeavours.

HOOKE: But you used *my* theory of Gravity. So the least you could do in your *Principia* is to confess your debt to me.

NEWTON: I 'confess' nothing! You simply repeated the same unproven theorem that Wren and Halley have been vainly wrestling with for years; and which *I* have been fully familiar with for over *two decades. (Returning the book to the shelf)* As to your challenge—'to calculate the trajectory followed by the Earth... '

HOOKE: *(Chiming in)* '... Under the influence of the attractive force of the Sun; and prove that it varies as the Inverse Square of the Distance'.

NEWTON: I concede it was a task worthy of my talents. So I worked out the solution by my method of infinitesimals, and calculated the trajectory to be... *(Describing with his hands)* ... an *Ellipse.*

HOOKE: As *I* predicted, Mr Newton.

NEWTON: Yes, but you were compelled to come to me, Mr Hooke, because *I* was the only man alive who had the mathematics to *prove* it.

HOOKE: But what of *my* Inverse Square Law?

NEWTON: The Inverse Square Law is but a consequence of Kepler's Laws, plus my equation for Centrifugal Force. And by combining these, with the help of my Fluxions, I have successfully calculated the elliptical orbits of *all* the planets around the Sun.

HOOKE: I still affirm...

NEWTON: *(Overriding him)* As a result, I have deduced that this gravitational relationship applies more *generally* than just to the Sun holding the planets in their orbits. The same Force holds *us* on the Earth. So there is a *Universal* Force—whereby every object pulls on, or is attracted to—every *other* object. And strange though it may seem, even you and I are attracted, my dear sir. But I knew all this five years ago, and made the appropriate calculations at the time.

HOOKE: Then why have you waited so long to publish these wonders?

NEWTON: As I told Mr Halley when he visited me in Cambridge, I had misplaced the relevant papers.

HOOKE: But *before* the publication of your *Principia,* you had published *nothing* for over *twenty* years! So why should anyone believe *your* dubious chronology of these events?

NEWTON: Because it is true.

HOOKE: And the Calculus? Or your Fluxions, as you call 'em. Do you still insist that *you* invented them, despite Leibniz's prior publication of *his* Calculus?

(HOOKE passes a volume of Leibniz's work to NEWTON.)

NEWTON: Yes! What's more, I proved to that German plagiariser that *I* was the inventor. (*Slamming Leibniz's* Calculus *down on the table*) But he still failed to acknowledge it.

HOOKE: In much the same way, as when you suddenly 'found' all your 'misplaced' documents and published them in your *Principia...* (*Slamming the* Principia *down on the table*)... you 'failed' to mention your incalculable debt to *me*!

NEWTON: I have acknowledged all my debts to such giants as Galileo, Copernicus, and Kepler. That is more than sufficient.

HOOKE: The arrogance of the Christmas Day Progeny knows no bounds, does it?

(*Smiling, NEWTON opens the* Principia.)

NEWTON: 'Nearer the gods no mortal may approach'.—Mr Halley's words, not mine.

HOOKE: It must be a wonderful thing having the Almighty do all your work for you.

NEWTON: It's true that God was by my side. But *I* had to do the work. In order to complete my labours, I never strayed from my rooms for almost two years.

HOOKE: Yet despite the praise you've received from the half-dozen worthies who have had the inordinate patience to struggle through your *Principia*, your publisher tells me that hardly anyone else has bought it. Perhaps this is because it is so abstruse.

NEWTON: I made it abstruse, sir, intentionally... (*Pointedly*)... to avoid being bated by little smatterers in mathematics.

HOOKE: Its impenetrable abstruseness is also why you have so few students. I have heard that when you lecture from your *Principia* at Trinity College, you spout to empty walls.

NEWTON: For years I have 'spouted' thus, sir. (*Amused*) And the occasional student that *does* attend my lectures is totally confused. Last week I passed a perplexed youth down by the Cam who muttered at me, 'There goes a man that writ a book that he nor anyone else understands'.

HOOKE: You find that droll?

NEWTON: *(Laughing)* Certainly. Having changed Man's conception of the Universe, I am proud that my achievements are only comprehended by the finest minds of the age. I am like the Alps, Mr Hooke. My peers will either be forced to climb me, or they will have to go around me. No scientist will ever be able to ignore me again.

HOOKE: So you really do believe that you are 'the most perfect mechanic of all'?

(NEWTON hands HOOKE the Principia.*)*

NEWTON: Can you refute any of my Laws? Or disprove my deductions concerning the motions of every object in the Universe?

HOOKE: I take issue more with the originality of your Laws. Your First Law is but a restatement of the laws proposed by Galileo and Descartes. As for your Second Law, that... *(Reading from the* Principia*)* ...'the acceleration of a body is proportional to the Force applied to it, and is inversely proportional to its Mass'...

NEWTON: *(Interrupting)* 'Mass' is yet another of *my* additions to the lexicon of science, Mr Hooke; and has nothing to do with Galileo or Descartes. For 'Mass is the material property of an object that denotes the quantity of particles within in it'...

HOOKE: Yes, yes, I know all this, Mr Newton. I'm not one of your many absent students.

NEWTON: Then I challenge you to tell me, Mr Hooke, when I release this apple—so... *(NEWTON drops the apple)*—why is it compelled to the ground?

HOOKE: Because of the gravitational attraction between the masses of the apple and the Earth, of course.

NEWTON: But the Earth's mass is composed of millions upon millions of particles which are all at *different* distances from the apple, and they pull on the apple accordingly—with lesser or greater ferocity.

(HOOKE revolves the globe on the table.)

HOOKE: Yes, but because the Earth is a *sphere*, there is as much mass on the Eastern side of the globe as there is on the Western side. So the *sideways* gravitational pull from both sides exactly *cancel* each other out—and the apple falls vertically towards the earth.

NEWTON: The fall is vertical, I grant you. Yet the problem is far from solved. For the same cannot be said for the forces exerted on the apple by the Northern and Southern hemispheres. The mass of the Northern hemisphere is, of course, *closest* to the apple. *(NEWTON points to the top of the globe)* Consequently it pulls with much greater ferocity than the Southern hemisphere which is much *further* away. *(NEWTON indicates the underside of the globe)* So... what is the ferocity—or the over-all magnitude—of the Earth's gravitational pull upon the apple?

(NEWTON picks up the apple and drops it.)

HOOKE: If I had the time, sir, I would humour you in this, too, but I know that you're aching to enlighten me. So please do so.

NEWTON: The answer is...

HOOKE: Yes, Mr Newton?

NEWTON: ...In my book, Mr Hooke.

HOOKE: You are the most infuriating man in Christendom!

NEWTON: But as you refuse to *read* my book, Mr Hooke, because the truth makes you bilious, allow me to enlighten your darkness. I have mathematically proved that both the *direction* and the *magnitude* of the Earth's gravitational pull upon the apple is as if *all* the particles of the Earth were *concentrated* at the Earth's *precise centre*. And that is a miraculous answer, as I'm sure you will agree.

HOOKE: God in Heaven, if you are right, then every object in the Universe, no matter its shape or size, can be now treated as if *all* its mass was concentrated at this *centre* of mass?

NEWTON: Exactly. Which brings us to my Third and most original Law that states…

HOOKE: 'That to every action there is always opposed an *equal* and *opposite re*action'. So?

NEWTON: So… I have used this Law to explain the *inter*-relationship of *all* phenomena in the Universe. Ergo: the Moon pulls on the Earth with the *same* force as the Earth pulls on the Moon. The same is true with the apple and the Earth.

(HOOKE laughs as he picks up NEWTON's apple.)

HOOKE: Are you suggesting that your legendary apple *pulls* on the Earth with the *same* force as the Earth pulls on the apple?

NEWTON: Yes.

(HOOKE drops the apple.)

HOOKE: *(Ironic)* And obviously because the apple is so 'powerful', the apple is *compelled* to *fall* to the *Earth*.

NEWTON: Exactly. The Earth's forces—that are exerted upon the apple—cause the apple's position to change *visibly*. Whereas the Earth, because of its immensely *greater* mass, *seems* totally unaffected by the apple. Even though it is not.

HOOKE: Hm!

NEWTON: *(Amused)* So there is no hiding from this Gravitational Force, Mr Hooke. Every atom of every object in the Universe is attracted to every other atom in the Universe, no matter how distant. And I have precisely defined—for the first time—the concepts of Mass *and* Force, and set down their relation to an object's Position and Acceleration. *(Smiling)* So with my minuscule contribution to Science, I have proved that my Theory of Universal Gravity must apply from the smallest grains of sand, to the Sun and the Earth, and to every planet, moon, comet, and star in the cosmos.

HOOKE: Your genius is undeniable. But don't you find it a terrible thing that you have not an atom of generosity in your soul?

NEWTON: Why should I? What's more, if you persist in publicly accusing me of plagiarism, I will hound you to your grave.

HOOKE: I believe you will try. But the battle is far from over.

NEWTON: To the victor all the spoils, then. The name of the vanquished is writ in sand.

HOOKE: And *you* are the sea, I presume.

NEWTON: Thou sayest.

HOOKE: There is no reasoning with you.

NEWTON: With me, there is *only* reason.

HOOKE: Then, God help us both.

(HOOKE leaves. NEWTON opens his Principia. *FATIO appears from behind the shelves.)*

FATIO: Your pardon, monsieur, but I happened to overhear your conference with that odious fellow, Hooke.

(NEWTON is disconcerted that FATIO looks like WICKINS, except that the dandified FATIO is beardless.)

NEWTON: It's not possible.

FATIO: What, monsieur?

NEWTON: You have the eyes of...

FATIO: Of whom, monsieur?

NEWTON: A firedrake comet I once knew.

FATIO: *(Laughing)* I have been called many things in my life, Monsieur Newton, but never 'a firedrake comet'.

NEWTON: Why have you appeared now?   I have grown used to walking alone in the abyss of space.

FATIO: That is one of the many reasons I have presumed to intrude upon your privacy. You see, I have read the first two Books of your masterpiece. And I here affirm that they are only one step removed from the sanctity of the Bible itself.

NEWTON: That is most flattering, sir, but bordering on blasphemy.

FATIO: It is as true as the Gospels. But there are still many questions that I would be grateful if you would deign to answer.

NEWTON: Which are?

FATIO: Because of their complexity, monsieur, perhaps I could induce you to dine with me at *The Cheshire Cheese*, where - over a tankard of porter - you could enlighten my darkness.

NEWTON: You are very persuasive, sir, and seem of good breeding but…

FATIO: I wish to be the acolyte of the greatest Magus of this age, as surely as my name is Nicholas Fatio de Duillier. *(Looking NEWTON in the eyes)* And there is no limit to what I will do to assist you.

NEWTON: *(Amused)* Well… let us to *The Cheshire Cheese*, Monsieur Fatio de Duillier, and we will see… what we will see.

FATIO: Please call me 'Nicholas'.

(NEWTON laughs at FATIO's effrontery. They leave the bookshop together… as THE LIGHTS FADE.)

Act 2, Scene 2

NEWTON's room/laboratory. Trinity College. Cambridge. Five years later. August. 1693.

The voices of NEWTON and FATIO can be heard in the passageway.

NEWTON: *(Off)* I've been feverish with worry, Nicholas.

FATIO: *(Off)* Did you not receive my letters?

NEWTON: *(Off)* Yes. Didn't you receive mine?

FATIO: *(Off)* Yes. Please stop fussing. I am weary.

NEWTON: *(Off)* I'm sorry. Where is your luggage?

(FATIO enters, taking off his silver-grey cloak as he does so.)

FATIO: At *The Three Bells.*

NEWTON: *(Following him in)* Surely you are staying here with me?

FATIO: *(Fingering some fruit)* Is there nothing to eat save these shrivelled apricots?

NEWTON: I hadn't noticed that they were shrivelled.

FATIO: Ugh. This cheese is not even fit for luring mice.

NEWTON: Again I...

FATIO: ...Hadn't noticed. *(Pouring himself some wine from an open bottle)* I assume your wine is sour, too.

NEWTON: For pity's sake, Nicholas, you promised...

FATIO: *(Interrupting)* You are forgetting that I suffered with a vomica upon the lungs.

NEWTON: No! When you wrote me of your illness, I was at my wits' end with terror. Didn't the money I sent you arrive?

FATIO: Open a new bottle. This is like imbibing phlegm.

(NEWTON takes a bottle from a cupboard and busies himself with opening it.)

NEWTON: So why won't you stay here with me?

FATIO: I must return to London tomorrow.

NEWTON: Now your health's improved, surely you can stay here a little while?

FATIO: *(Overriding him)* Give the bottle to me. You are all thumbs and fingers.

NEWTON: Is it surprising when you won't tell me *why* you're not staying?

(FATIO takes the bottle from NEWTON and uncorks it.)

FATIO: I have some business in London that compels me to return.

NEWTON: And the nature of this 'business'?

(FATIO pours himself some wine.)

FATIO: I have been working on a process to purify mercury with a friend.

NEWTON: What friend?

FATIO: Together we took some filings of gold, and made a softish amalgam of them. Then we put the amalgam in a sealed egg, and the sand heat made it presently puff up.

NEWTON: Don't trifle with me. Tell me who this friend is!

FATIO: Then the amalgam grew black, and passed through the colours of the Philosophers in a matter of seven days.

NEWTON: And *nights*, I've no doubt!

FATIO: Then there grew a heap of trees from the matter, which changed by degrees their colours, beginning near the base with verdigris and violet....

NEWTON: *(Interrupting)* Why tell me what I know? I recorded this process in my *Clavis* over two decades ago. But this new friend of yours; I don't know *his* name, so what is it?!

FATIO: It is unimportant.

NEWTON: Not to me! And I can see that he is of more than pressing importance to you.

FATIO: Is it surprising? It was my friend's secret remedy that cured my illness. As a result, I intend to become a physician myself which will not require above a year or two of my time.

NEWTON: Another of your skittish fads.

FATIO: No! I am going to enter into a partnership with the creator of this life-giving elixir. Then I will cure thousands of people at little cost, and thereby raise a fortune for myself.

NEWTON: And the catch?

FATIO: The cordial's drawback is its emetic qualities, but I am sure that we can overcome the excessive watering of the stools.

NEWTON: So I repeat, Nicholas; what is the catch?

FATIO: Naturally I must expend about 150 pounds a year for about four years until the elixir has established itself.

NEWTON: Ah.

FATIO: But I would never propose to your good self to be a partner in this venture because of the possible accidents that could befall the enterprise.

NEWTON: I think I will have some of my sour wine.

FATIO: What is the alternative to my selling the elixir? Now that my meagre inheritance has dwindled to nothing.

NEWTON: *(Pouring some wine)* You can live with me and...

FATIO: '...Be your love / And we will all the pleasures prove'.?

NEWTON: *(Spilling some wine in his anger)* Don't quote that atheist in my presence!

FATIO: *(Laughing)* That is not what frightens you about Christopher Marlowe.

NEWTON: You find our friendship laughable, Nicholas?

FATIO: No. But I wish there had been *more* laughter between us. Then there might have been a chance...

(FATIO trails off.)

NEWTON: Of what?!

FATIO: Why are you so choleric?

NEWTON: You know only too well that your dubious venture with this 'friend' of yours, is the ultimate betrayal of our... our...

FATIO: Yes?

NEWTON: ...Intimate feelings for one another! I'm not frightened to say it.

FATIO: We have gone beyond that now.

NEWTON: Why?

FATIO: I know how you refer to me with your friends. That's why they call me 'Newton's Ape'.

NEWTON: Nicholas, you mustn't listen to salacious tongues.

FATIO: And that is not the worst of it.

NEWTON: How else have I offended you?

FATIO: You will agree that I have corresponded, as an equal, with the world's elite, including Huygens, Leibniz…

NEWTON: Don't speak of him!

FATIO: …And Domenico Cassini.

NEWTON: Yes, I know.

FATIO: *(Overriding him)* Yet despite my acknowledged brilliance, you still will not allow me to supervise the second edition of your *Principia,* when I am more than qualified to do so.

NEWTON: I am sorry, Nicholas, but your constant bragging prompts misgivings.

FATIO: *(Drinking)* That is what I mean.

NEWTON: It's only the truth I speak, dear friend.

FATIO: *(Shaking his head)* You do not understand, do you?

NEWTON: I understand that Huygens wrote to me, telling how you boasted to him that *your* edition of the *Principia* would be far longer than my original because you would add so much more to it.

(FATIO finishes his glass and pours himself some more.)

FATIO: You only listen to the surface of what is being said. It is true. For also I have produced a formidable theory to *explain* Gravity…

(FATIO trails off in response to NEWTON's harsh laughter.)

FATIO: You see. My only reward as your acolyte is barking derision.

NEWTON: I wouldn't jibe, Nicholas, if you would only refrain from making assertions that you can't possibly substantiate.

FATIO: Such as?

NEWTON: You have just claimed that you can *explain* Gravity, when even *I* acknowledge that I cannot. It is one thing to provide a mathematical *description* of Gravity, but to explain its true physical *cause* is quite beyond the wit of man. And certainly beyond yours.

FATIO: I always wondered why Wickins has had nothing to do with you ever since he left you. Well, he was your first 'firedrake comet', was he not?

NEWTON: That's not how I see you. Your eyes and face are as dear and clear to me...

FATIO: *(Completing his sentence)* ...As moon-washed merd!

NEWTON: You can't leave me like this.

FATIO: No man can abide with you.

NEWTON: There is nothing I wouldn't do for you, Nicholas. And you know that.

(FATIO pours himself the last of the wine.)

FATIO: So you are willing to furnish my friend and me with the necessary funds to propagate our elixir, are you?

NEWTON: Who is he? I *will* have his name, God damn you!

FATIO: *Now* who is the blasphemer?

NEWTON: You cannot treat me so!

(FATIO rises and puts on his cloak.)

FATIO: What will you do? Pursue me to London and give vent to your jealous fury in the Royal Society, and delight your enemies, Hooke and Leibniz?

NEWTON: Why do you say such things when you know your words are envenomed sword-tips in my brain?

FATIO: Because every other man but you would have said they were stabbing his *heart.*

(NEWTON takes another bottle of wine out of the cupboard.)

NEWTON: Take more wine, Nicholas. It'll help temper your caustic tongue.

FATIO: Your wine is pig-swill, sir. Out of my way.

(NEWTON falls to his knees, clutching FATIO's cloak.)

FATIO: Whatever are you doing?

NEWTON: I won't rise from my knees until you stop your cruelty.

FATIO: So it has come to this.

NEWTON: Even if you no longer love me, surely you will not leave me bound upon the wheel of such exquisite torment?

FATIO: For pity's sake, get to your feet, man. Grovelling ill becomes the Scientific Colossus of the age.

NEWTON: *(Rising)* See? I'm standing as you wish. Now promise me you'll stay.

FATIO: How can I stay with someone whose hunger for position and power exceeds his love of his fellow man?

NEWTON: You chastise me unfairly, Nicholas. When I resigned my Lucasian Chair, I was a Member of Parliament for Cambridge only one paltry year.

FATIO: But ever since you have daily dog-licked the Earl of Halifax, begging him to bestow upon you some other 'imposing position of authority'. And you do not care what that position is as long as you can glow within its golden penumbra.

NEWTON: You wrong me in every way!

FATIO: You are a scrivening coward, sir.

NEWTON: You exceed all bounds!

FATIO: Do you deny that you hypocritically attacked Boyle for publishing his daring thesis on Alchemy, while you were secretly still continuing your own quest for the Philosopher's Stone?

NEWTON: It would be arrant foolishness to *publicly* reveal...

FATIO: *(Overriding him)* And you are also utterly craven when it comes to upholding your Religion!

NEWTON: You must understand my position.

FATIO: Oh I do, I do. But I am sick to my bowels with your morbid secrecy, so I'm going to London this very night.

NEWTON: Nicholas, despite your calumny, I have never loved man or woman as I have loved you.

FATIO: I loved you, too, Isaac.

NEWTON: But I love you *still*!

FATIO: You love with the voraciousness of a vampire! Your jealousy knows no end. Oh I will always defend you to the world. But I cannot go on living in your shadow.

NEWTON: I'll suppress my feelings, dear friend, and give you greater freedom.

FATIO: These are vain promises, Isaac. And in the depths of your feverish mind, you must know that a love such as ours, if it were ever discovered—or even guessed at—would spell your ruin.

NEWTON: *(Ironic)* So—by deserting me, you are saving me from myself?

FATIO: View it how you will, I must attend the London coach.

NEWTON: Dearest friend, you are the only human happiness I have ever known.

(NEWTON embraces FATIO.)

NEWTON: I cannot—and I will not—let you go!

(FATIO frees himself from NEWTON.)

FATIO: And I cannot tolerate one more day being called 'Newton's Ape'. As the Lord is my witness, no one will call me that again.

NEWTON: *(Clutching his elbows)* I beg you, Nicholas, please don't leave me!

FATIO: *(Shaking NEWTON off)* I must. Now do not pursue me to the street, or I will not be answerable for what I do.

(FATIO leaves, slamming the door behind him. For a moment NEWTON stands dazed, then he emits an anguished wolf's howl. Using his nails like talons, NEWTON rips one of his crimson cushions apart. Like a dervish he flails amid the swirling feathers, throwing books and furniture about him in a frenzied storm.

As suddenly as it began, NEWTON's fury subsides. With tears streaming down his face, he is now on all fours like a beaten animal amid the debris of his room. His posture is reminiscent of William Blake's famous engraving of NEWTON.

There is a knock on the door but NEWTON is too preoccupied with his desolation to register it. There is another knock but still he does not respond. As the door opens, NEWTON's tear-stained face turns towards it.

CHARLES MONTAGU, First Earl of Halifax, appears in the doorway. [He is played by the same actor who portrayed FATIO.] But unlike FATIO, who sported a silver-grey cloak and was hatless, MONTAGU wears a midnight-blue cloak and his large black hat obscures the upper part of his face.

Because of NEWTON's deranged state, NEWTON scrambles to his feet and embraces the disconcerted MONTAGU, believing him to be FATIO.)

NEWTON: I knew you would not forsake me.

MONTAGU: What's wrong, Isaac?

NEWTON: How could you say that your face was no more to me than moon-washed...?

(NEWTON trails off.)

MONTAGU: Isaac, don't you... know who I am?

NEWTON: *(Slowly realising that he is not addressing FATIO)* Yes... yes... I see now that you are not as I... supposed. Your face is... similar - but your eyes... *(Laughing distracted)* Even they have something in common. But then, to me, all men's faces tend to merge into one face. What goes on *behind* their eyes is much the same.

(NEWTON lurches away from MONTAGU.)

NEWTON: But why should I expect anything different? From time immemorial the Earth has been peopled with whoremongers copulating like butterflies!

MONTAGU: My dear friend, are your drunk?

(MONTAGU tries to steady NEWTON who is swaying unsteadily.)

NEWTON: Take your hands off me, Montagu! Lechery contaminates.

MONTAGU: Who has done this to you? *(MONTAGU surveys the wrecked room)* Tell me and I will have the Watch apprehend your assailant.

NEWTON: *(Blankly)* *Everyone* assails, Montagu; but you cannot apprehend the whole world.

MONTAGU: For mercy's sake, my friend, sit down...

(MONTAGU tries to help NEWTON into a chair but NEWTON shrugs him off.)

NEWTON: Locke boasts that he is a Philosopher King, and Wren proclaims that he has designed the greatest dome in Europe. But the damnable truth is that they are both whoresons who wish me riddled with the pox, like *they* are!

MONTAGU: You speak nothing but filth, Isaac. They are amongst the most honourable men in the kingdom.

NEWTON: *You* are no better, my Lord Halifax. I've seen you squinney at my niece, and watched her simper her moist lips at you.

MONTAGU: You do not know what you are saying, sir.

NEWTON: Even the motes of dust have rutting microbes on their backs. *(Moving to the open door)* The particles in the rainbow are as heinous.

MONTAGU: Where are you going, Isaac?

NEWTON: To find the solace of an apple tree down by the Cam.

MONTAGU: You can't walk the river-bank in your delirium for fear...

NEWTON: *(Overriding him)* I fear no man, sir! But I fear *for* all men. The shame is, so few realise that chastity is the only panacea.

MONTAGU: I came here with such news, Isaac, to make you smile.

NEWTON: There's nothing to smile at until Time heals. *If* Time heals...

(NEWTON shakes off MONTAGU's consoling hand.)

NEWTON: No, my Lord, your touch pollutes. Besides, I'm done with men – and women, too. Myself, alone, is all in all.

(NEWTON gestures at his upturned room.)

NEWTON: This room mirrors unrequited love... and that gives birth to violence.

MONTAGU: I must escort you for own good.

NEWTON: You are an eminent politician, aren't you, Charles?

MONTAGU: Yes, and once we Whigs come to power, I will be Chancellor. That's why I've come to see you today.

NEWTON: Then *act* the politician, my lord. *(NEWTON gestures again at his upturned room)* And bring order into chaos.

(NEWTON is now in the open doorway.)

MONTAGU: Isaac, are you certain that you are safe alone?

NEWTON: No man is safe within the fearful confines of his own skull.

(NEWTON leaves. FADE... on MONTAGU, shaking his head in disbelief.)

Act 2, Scene 3

A Cheapside brothel. London. Five years later. December. 1698.

The tawdry room has a dishevelled bed, two chairs and a table, with an open wine bottle and glasses. The door is ajar. A WOMAN laughs in the passageway.

WOMAN: *(Off, with a Cockney accent)* So why don't you come right in, my fine gentleman? Then you can more than just ogle me bubbies. You can thrust yer lusty tongue between 'em. *(Pause)* Oh—so that ent to yer taste. So wot you come here for if it ent to sample me ample wares?

(A HOODED FIGURE comes into the room, followed by ANNIE, who is played by the same actress who portrayed GRACE. But unlike the red-headed Yorkshire GRACE who was elegant and well-dressed, ANNIE is a poorly-dressed East End, raven-haired girl who is down on her luck, and has to 'expose her charms' in order to make a living.)

FIGURE: I've come to ask you some questions. Now sit down, and answer me.

(ANNIE lies back on the bed, displaying her 'wares'.)

ANNIE: *This* the kind of sitting you have in mind, saucy sir?

FIGURE: Don't speak until I bid you, or it'll be the worse for you.

ANNIE: Stop trying to frighten me, and take off yer hood like a true gallant.

(ANNIE undulates towards him.)

ANNIE: Or do you want to stay hooded 'till I've unsheathed your monster manikin?

(ANNIE starts to undo the FIGURE's belt but he pushes her away.)

FIGURE: Take your pox-ridden hands off me! Your name is Annie Limlet, isn't it?

ANNIE: *(Suddenly wary)* How'd you know that? *(Peering at him)* Who… are you?

FIGURE: Where's your pimp-master, William Chaloner?

ANNIE: Why won't you look me in the eyes like a proper man?

FIGURE: Where is Chaloner?

ANNIE: I know you!

(ANNIE rushes across the room and pulls the FIGURE's hood back to reveal... NEWTON.)

ANNIE: You're Newton, the Warden of the Mint!

NEWTON: Your careless tongue will drag you off to Newgate.

ANNIE: Now I see why Bill calls you 'that old hanging dog'.

(NEWTON pins ANNIE down on the bed and is about to strike her.)

NEWTON: One word more, you whore, and, so help me…!

ANNIE: …You'll beat me black and blue? That's the only way your sort can get yer droopy pizzles to spend yer spurt, ent it? Coz all the women of yer *own* sort treat you like puke, don't they?

(NEWTON recoils from her.)

NEWTON: I have no truck with women of *any* sort, you jade. You're no better than my mother was!

ANNIE: So your mother was a whore, too, was she?

NEWTON: No! I've had so many hours now without sleep, my tongue plays tricks.

ANNIE: *(Winking)* There are tricks and tricks, Warden.

NEWTON: I'll not countenance you soiling her name. I am here to apprehend your Chaloner for counterfeiting. And I will not rest until he swings from the gallows tree like the maggot-infested apple that he is. So where is he?

ANNIE: *(Laughing)* Why didn't you say all that from the start, Warden?

NEWTON: *(Surprised)* Do *you* also want to see the forger dangling from a noose?

ANNIE: I do! Well, Chaloner promised me that he'd take proper care of me as his Phillis. Instead when he had to flee your agents, he set me up whoring here, didn't he? Then once he were safe in the countryside—quick as a farmer drops his breeches at the sight of a sheep's arse—the dirty ram tupped every slut he could find from Camberwell to Kent. And he says he'll be *here* soon after sunset.

NEWTON: Excellent.

ANNIE: He's a violent man if surprised.

NEWTON: My agents will apprehend him on sight.

(There is a knock on the door.)

NEWTON: Seems they've caught the wretch already. Come!

(The door opens to reveal MONTAGU, swathed in a dun-coloured cloak.)

NEWTON: Charles, what ever are you doing in this carnal pit?

MONTAGU: Under the circumstances, Isaac, I've little choice but to ask the same of you. That's why I followed you here.

NEWTON: Wait in the passage way, Limlet. You eavesdrop at your peril.

ANNIE: *(With a mock curtsey)* To hear is to obey, O Mighty Warden.

(Giggling, ANNIE undulates out of the room.)

NEWTON: Have you taken leave of your senses, my lord?

MONTAGU: It is *you*, Isaac, who have taken leave of *your* senses coming to this stew. This is very like your previous brain distemper.

NEWTON: That temporary aberration is five years' past. I'm as settled in my faculties as the sanest in the realm. Besides, unlike you, my lord, I am here officially.

MONTAGU: Oh come now, you cannot claim that you are 'officially' in a clap-ridden brothel!

NEWTON: Well, I'm certainly not here to contract the pox, sir! I have laid a trap to ensnare one of the most cunning counterfeiters of the age, but it will come to little if he espies a meddlesome mouse.

MONTAGU: And you call that thanks for all that I have done for you?

NEWTON: I am deeply grateful, my lord, that I was appointed Warden at your instigation. But, in recompense, I have over-seen the entire re-coinage of England. It's only because of *my* exhaustive calculations that the movement of the coining press and the action of the coiner are now so co-ordinated that 2 mills with 4 millers, 12 horses, 3 cutters, 2 flatters, 8 sizers, 1 nealer, 3 blanchers, 2 markers, 2 presses with 14 labourers to pull 'em, can coin at the rate of 3000 pounds sterling per diem!

MONTAGU: Calmly, Isaac, calmly.

NEWTON: As a result, the Mint has made six and a half million new pounds in three years which is a gargantuan achievement, considering barely half this amount was produced in the previous thirty years.

MONTAGU: Such frenzy could trigger off another brain seizure.

NEWTON: Is my frenzy surprising? I'm utterly exhausted pursuing and hanging every counterfeiter that stalks the land. For if I fail in my mission—and no one can be more aware of this than *you* as Chancellor—our economy will break like a rotten branch in a storm. Then there'll be a social upheaval comparable to the Civil War. Now leave me, sir, so I can go about the Queen's business.

MONTAGU: As you will. Though I dearly wish you wouldn't always pursue these dangerous miscreants in person.

NEWTON: I've never left to others what I can do myself.

MONTAGU: Even though many hate you for it, and call you the Gallows Warden?

NEWTON: Even so. For many more would hate me if the economy juddered into ruins. And before you moralise further, my lord, you should look to your *own* conduct.

MONTAGU: What do you mean?

NEWTON: Don't play the innocent with me, Charles. It doesn't become a man of your pre-eminence. *(Pointedly)* My niece, sir.

(MONTAGU moves to the door.)

MONTAGU: As you say, Isaac, it's best if I leave you to your mission.

(MONTAGU closes the door behind him. The light is beginning to fail. NEWTON takes out his fob watch. The door opens to reveal ANNIE, with a lighted candle.)

NEWTON: Wench, I told you to wait for my...

(NEWTON trails off when he realises that ANNIE is not alone. Behind her is a hunched FIGURE.)

NEWTON: The house is surrounded, Chaloner. If you raise a hand against me, my agents will shoot you down like the vermin that you are.

(The FIGURE laughs, then hobbles into the room with his stick. It is HOOKE, who looks frailer than ever. HOOKE laughs and gestures towards ANNIE's charms.)

HOOKE: Tut, tut, Mr Newton. I always thought you had the soul of a fastidious Puritan. But then perhaps you do. It's just your loins that aren't so fastidious.

NEWTON: So this is where you frequent? I should have surmised as much.

(HOOKE points to ANNIE.)

HOOKE: *(Amused)* Have you enjoyed her? If not, we can always share her. Well, you once ogled my niece Grace, and despite Annie's tawdry attire, there is a definite resemblance.

NEWTON: I'm on Her Majesty's business and I will not be provoked.

HOOKE : Leave us, Annie. I'll come and pleasure you by and by.

NEWTON: Limlet, you know the penalty for listening at doors.

(Laughing, ANNIE closes the door behind her.)

NEWTON: As for you, Mr Hooke, you must return tomorrow to assuage your carnal anguish. *I* require the wench's services tonight.

HOOKE: *(Laughing)* Of that, I have no doubt.

NEWTON: It is not as your soiled morality supposes.

HOOKE: I believe you.

NEWTON: You do?

HOOKE: *(Smiling)* Yes. If you *are* capable of being moved by human desire, I'm sure your eye squints quite another way.

NEWTON: I have no time for verbal fencing, sir. I'm here, as Her Majesty's Warden of the Royal Mint, to apprehend a forger, so I must insist that you leave now.

(HOOKE pours some wine into a glass.)

NEWTON: *(In disbelief)* Didn't you hear me, sir?

HOOKE: Yes, but I'm unaccountably dry, sir. Would you care for a bumper?

NEWTON: No!

HOOKE: I'm glad to see you no longer have your ape upon your shoulder.

NEWTON: I don't wish to speak of Fatio.

HOOKE: In spite of Fatio's passionate defence of your *in*defensible position *vis à vis* Leibniz and *his Calculus*?

NEWTON: What has Leibniz to do with Fatio?

HOOKE: So you have not read your progeny's defence of you in his grandiose 'Lineae Brevissimi Descensus Investigatio Geometrica Duplex'?

NEWTON: No, I have not!

HOOKE: Then the rumours that you and your anthropoid parted abruptly are true?

NEWTON: They lie!

HOOKE: So why did your mental distemper follow so fast upon the heels of Fatio deserting you?

NEWTON: More lies!

HOOKE: But that's not the worst that has befallen you, is it?

NEWTON: This is the last time I shall say this; I am here, on Her Majesty's business, which *you* are severely hindering.

HOOKE: My very point! In order to keep your post as Warden of the Mint, you are being forced to pay an ignominious price.

NEWTON: What price, sir?

HOOKE: The price of having his lordship, the Chancellor—who I was honoured to pass just now upon the stairs—always under your roof, so he can continue to make the beast with two backs with your delectable niece, Catherine.

NEWTON: That is mendacious filth!

HOOKE: And though it appals you to the depths of your Puritan soul, you still dare not challenge their rabid lust because you fear the Chancellor will have you replaced as Warden of the Mint.

NEWTON: Take back that slander or I will beat you like a cur!

(NEWTON threatens HOOKE with his stick.)

HOOKE: Why pause, O Mighty Warden? From the moment you boasted that you have 'seen further by standing on my giant's shoulders', you've dreamed of flagellating my deformity. So why stay your hangman's hand now?

NEWTON: *(Lowering the stick)* I do not know, sir... but it were good you left me to my mission.

HOOKE: Don't you see the irony of where you have arrived, Isaac?

NEWTON: For your health's sake, go! My mission demands...

HOOKE: ...That it's more important for you to send poor folks to the gallows than it is for you to pursue scientific truth - which *was* the mission of your life. But now Natural Philosophy is dead to you. You have driven a stake through its heart.

NEWTON: I will not countenance you defaming my good name.

HOOKE: Even your friends say that you feed the executioner by day, and then come the night, you spend fruitless hours with arcane texts, foolishly trying to reconstruct Solomon's Temple in your library. Oh Isaac, Isaac – how are the Mighty fallen!

(Closing the door behind him, HOOKE leaves the anguished NEWTON to his mission as... THE LIGHT FADES.)

Act 2, Scene 4

HOOKE's chambers. Gresham College. London. Five years later. March. 1703.

In addition to the usual furniture, there is a four-poster bed with the curtains drawn around it. Behind the curtains there is asthmatic breathing.

ANNIE LIMLET, who is now slatternly, comes into the room, carrying a bowl of gruel.

ANNIE: No wonder you're dyin', Robby. When you do nuffink but fill yer sickly guts up wiv senna, rhubarb, wormwood, laudanum, flowers of sulphur, resin of jalop, Aldersgate cordial, Epsom water and 'agiox. And that's just before breakfast.

(ANNIE pulls open the curtains around the four-poster bed to reveal the half-blind and ailing figure of HOOKE.)

HOOKE: I don't have to be abused in this way. Give me my gruel.

ANNIE: And no wonder you're half blind, when you make me blow that powdered human dung in yer eyes to take off its films.

(HOOKE snatches the gruel from ANNIE.)

HOOKE: *(Coughing)* Stop playing the scald.

ANNIE: An' how can you possibly believe that the blood of a black cat will cure yer chilblains?

HOOKE: Have you been reading my diary, hussie?

ANNIE: Yes, and I always thought you were a droolin' lecher. Then once I'd worked out wot all them numbers in yer diary meant, they proved it.

HOOKE: What numbers, wench?

ANNIE: Them numbers that you always wrote after you'd pronged that whore, Grace, and all yer other housekeepers.

HOOKE: Grace was no whore. I truly loved her. For she was...

ANNIE: ...Yer niece who you straddled like any common tart. And them numbers tell how powerful your spunk was, don't they?

HOOKE: It takes a whore to know one! And that's what *you* were, Annie Limlet, 'till I rescued you from your bawdy house.

ANNIE: You call livin' in this dung hole, with a stinking, old goat like you; 'being rescued'?

HOOKE: Oh be off with you, you harridan!

ANNIE: *(Wrenching open the door)* I'll be off for good and all, when you pay me wot you owe me, you miserly limp hook.

HOOKE: Where are you going?

ANNIE: To the gin shop. And don't expect me back 'till nightfall. If then!

HOOKE: You can't leave me unattended, Annie!

ANNIE: Just watch me.

(ANNIE exits. HOOKE coughs fresh blood into a rag and slumps back into his pillows. He is not aware of the door opening, or the appearance of NEWTON.)

HOOKE: If only the Almighty would bring some light into my darkness. *(Sensing he is not alone)* Who's there? No need to tell me. I know you. *(Chuckling)* The Angel of Death. Angel Isaac. Come to gloat, have you?

NEWTON: No. I heard that you were not your best.

HOOKE: If that's how you define dying, then you have it right. But don't worry, when I'm gone you will attain the position that you've coveted these many years.

NEWTON: I've no wish to don *your* mildewed mantle of 'Secretary of the Royal Society', if that's your meaning.

HOOKE: No, once I'm worm's meat, you will swiftly become the *President* of the Royal Society. Then, in the safe knowledge that you'll be spared my much-feared critiques, at last you'll be brave enough to publish your treatise on 'Optics'.

NEWTON: I fear no man's opinion.

HOOKE: *(Laughing)* You may fool you, Isaac, but you don't fool the Almighty. What's more, in His Infinite Wisdom, the Almighty may well send *me* back to haunt you - and thereby keep you honest.

NEWTON: I don't believe in wraiths.

HOOKE: I can see the incubi behind your eyes.

NEWTON: Don't try to demonise me, Hooke! Only God shines through me.

HOOKE: So clarions the Christmas Day Progeny to the last.

NEWTON: I'm the Lord's Instrument, nothing more.

HOOKE: Aren't you at least His Son?

NEWTON: How can you blaspheme, when you are on the verge...?

HOOKE: ...Of being dragged down into the bowels of Hell? For that is where those that embrace the Sin of Pride will meet – as *you* must know from having spent far more years on biblical studies than you ever did on Science. Though, it's true, the other day I re-read your *Principia*. And again I felt a sense of wondrous awe. Especially as it seemed to me that you had metamorphosed *us*... into the Sun and the Moon.

NEWTON: *(Laughing in disbelief)* Hooke, this is just another feverish delusion of...

HOOKE: ...A dying man? That being the case, I'm sure that even a rationalist like yourself will give ear to my 'feverish delusions'. For here on my death bed, like John of Gaunt, I prophecy that as the flux and reflux of the Sea arise from the gravitational actions of the Sun and the Moon; so future scientists will acknowledge that *you* as the Sun and *me* as the Moon, in our different ways, have exerted a combined Force upon the Sea of Natural Philosophy.

(NEWTON emits a barking laugh.)

NEWTON: My poor, demented fellow, scientists of the future will only acknowledge *my* achievements because I am both the Sun *and* the Moon. Whereas, at best, you are just a giant breaker that has endlessly spumed your froth against the unforgiving rocks. Yes, sadly you're a Renaissance man of the past, Robert, who has come too late into *my* world. For with my *Principia Mathematica,* I have ushered in the Age of Reason which is mankind's future.

HOOKE: You think I didn't know what I was doing when I challenged you, Isaac? I knew only too well that I was about to release the genie— and your genius—from the bottle. Then once you had mathematically described the Forces in the Universe, I was equally sure that the genie could never be returned to the bottle. But I challenged you because I believe that scientific truth must prevail, whatever the cost. The appalling irony is that someone as unforgiving and ungenerous as you are should prove to be the Way and the Light.

(HOOKE laughs and falls back against his blood-stained pillows.)

NEWTON: I'm glad you are capable of mirth at such a time.

HOOKE: *(Still laughing)* What better time than now? Well, I can foresee what you will do when you become the President of the Royal Society. As you want all the credit for yourself, you will have my portrait burnt, my inventions discarded, and my name all but obliterated.

NEWTON: What stuff and tush, man. Under your ailing tutelage as its Secretary, the Royal Society has fallen into such disarray that it will take me years—as its President—to restore its woeful finances to their former health, so I'll have no time to pursue your ghost.

HOOKE: You'll find the time! Human beings to you are merely satellites that orbit around your Sun. Once they displease you, or you can't control them, you consume them in a corona of fire. So I further prophecy that when you have acquired your longed-for knighthood—as you soon will—you'll turn your solar wrath upon Leibniz and all your *other* enemies until they are only ashes in the face of Heaven. For though you are the greatest natural philosopher who has yet bedazzled the Earth, tragically there is a stygian Darkness in your heart, compounded of fury and revenge. This Darkness is corrupting your incandescent Light like a malignant tumour in an angel's breast. But you could yet *save* yourself *from* yourself; if you would only re-dedicate yourself to the light of Science. You have so much to give, and you still have the time to destroy the spiritual cancer that is eating your soul before *(Pointing at his own withered frame)* you, too, are reduced to this!

(HOOKE's lava flow of prescience is stifled as his life's blood erupts from his lungs. HOOKE falls back dead. NEWTON stares at HOOKE in disbelief. Then NEWTON closes the dead man's eyes.)

NEWTON: Robert... Robert... at peace... *You* are at peace now. *(NEWTON places pennies over HOOKE's eyelids)* Whereas I—it seems —have only the future that I have created to look forward to. And without you to goad me into venturing beyond myself. Farewell, old adversary, your like will not come again.

(Hearing a noise, NEWTON pulls the four-poster curtains, concealing HOOKE, as a HOODED WOMAN in a green cloak appears in the doorway. Everything about her is reminiscent of the hooded CATHERINE, NEWTON's childhood sweetheart of Act 1, Scene 2.

(NEWTON backs away from the HOODED WOMAN.)

NEWTON: No, no... you died, murmuring my name. Or so they say.

(The WOMAN pulls back her hood and we see that the WOMAN is indeed being played by the same actress who portrayed CATHERINE. But unlike the Lincolnshire, flaxen-haired CATHERINE, the WOMAN has chestnut-coloured hair.)

WOMAN: Uncle Isaac, it's me, Catherine.

NEWTON: Catherine's dead. You're just a figment of my fevered brain.

CATHERINE: *(Laughing)* I'm no figment, Uncle. Or your childhood sweetheart, Catherine Bakon. Although you once said that my eyes reminded you of hers, when she was my age.

NEWTON: *(Laughing to cover his embarrassment)* Of course, of course, you're my niece, Catherine... Barton. Twilight has been playing tricks with my eyes again, and I *am*... infinitely tired. You have no idea how tired. Why are you here?

CATHERINE: I had strange misgivings so I followed you. Where is Mr Hooke?

NEWTON: In the deepest of sleeps.

CATHERINE: Come home with me, Uncle. I've made you some mutton broth that will restore colour to your cheeks.

NEWTON: *(In his own world)* If only there were an end to thinking! Never to be at rest. Never, ever.

CATHERINE: Let's go, Uncle, before we wake Mr Hooke.

NEWTON: There's little fear of that.

(CATHERINE takes his hand.)

CATHERINE: Your hand is very cold.

NEWTON: I have ice upon my heart, niece, that I fear will not melt in my lifetime. Yet that is nothing to the polar night that now embraces Robert Hooke. But in that night—God willing—there is a peace that passes all understanding. Though, in truth, Robert found a kind of peace *here*—in that he saw... sees each and every person—as an individual, and as a light in themselves. Whereas, tragically, I see more humanity in the mountains of the Moon than in mankind.

CATHERINE: That is a terrifying concept, Uncle, if you truly believe it.

NEWTON: I do—Heaven help me. But Physics should always beware Metaphysics.

CATHERINE: Why do you talk so distractedly?

NEWTON: Forgive me. I'm not myself tonight. And may never be again.

CATHERINE: Uncle, let me take you home. I have a great foreboding...

NEWTON: Until just now I truly thought I had discovered a great deal.

CATHERINE: You have! Sir Christopher Wren said that your rigorous principles of investigation have formed the foundation on which Science can flourish in the centuries to come. So how can you sit there in your dumps?

NEWTON: Because there is far, far more that I have *not* resolved.

CATHERINE: But you have determined the motions of the Sun, the Earth, the Moon, and planets, so there is *nothing* you have not discovered.

NEWTON: Yes, but that is very little when I have failed so miserably to understand my fellow men. And all because I have been blinded by the searing crucible of Science. But even there, although I have set out the motions of all the celestial bodies, I have far from determined the *remaining* structure of the Universe! The attraction of every Atom to every other Atom is merely the beginning. But the *structure* of the Atom - about which, as yet, we know nothing – *that* may be the answer to God's Cryptogram. When the Atom's structure is known, it could well blind us with truth like... the radiance of a thousand suns. *(Now standing by the window)* Look, moonrise over London.

CATHERINE: *(Tugging at his hand)* Uncle, for pity's sake.

NEWTON: Why is it that Nature does nothing in vain? Whence arises all this order and beauty that we see in the Universe?

CATHERINE: God's Riches are wondrous, but we must go.

NEWTON: In the light of the Almighty's Creation—what ever I may appear to the world—to myself I seem to have been only like a boy playing on the sea-shore, and diverting myself in now and then finding a smoother pebble, or a prettier shell... whilst the great ocean of Truth lies undiscovered... all before me...

(NEWTON gazes at the moon, then at the bed, and finally at his empty hands. Then he goes out into the night. CATHERINE stares after his hunched, retreating shadow, then she follows him out.

(The moon fills the cyclorama. Then it explodes like a thousand suns and we are left with the polar darkness.)

THE END

Second View: Three Minions

Calculus ("Newton's Whores")
by Carl Djerassi

(Time: 1712–1731. London,
mostly in a salon and/or sitting room)

CAST IN ORDER OF APPEARANCE

Colley Cibber (1671–1757), playwright, actor, theatre manager, eventually poet laureate (1730). Literary friend of Vanbrugh, literary enemy of Alexander Pope and John Arbuthnot. Author of *Love's Last Shift* (1696) and other plays. Completed Vanbrugh's *The Provok'd Husband* in 1728.

Sir John Vanbrugh (1664–1726), playwright, architect (of Castle Howard and Blenheim Palace), advisor to George I. Author of *"The Relapse: Or Virtue in Danger"* (1696), a highly successful sequel to Cibber's *"Love's Last Shift,"* as well as other plays. One of the first directors of the Royal Academy of Music.

Margaret Arbuthnot (?–1730), wife of John Arbuthnot, mother of six children.

Louis Frederick Bonet (1670–1762), citizen of Geneva, Minister of King of Prussia in London (1696–1719), then "syndic" and senator in Geneva. Trained in medicine and law, proselytizing Protestant. Fellow of the Royal Society in 1711, member of the Berlin Academy in 1713. Member of anonymous Royal Society Commission of 1712. *(This role can be played by the same actor as Colley Cibber.)*

John Arbuthnot (1667–1735), Scottish-born and Scottish-educated, physician to Queen Anne, some mathematical (statistical) knowledge, wit and satirical writer, friend of Pope, Swift, John Gay, and Thomas Parnell (founding member of Scriblerus Club in 1714). Author of the political allegory "History of John Bull" describing the prototypical Englishman. Fellow of the Royal Society in 1704. Member of anonymous Royal Society Commission of 1712. *(This role can be played by the same actor as John Vanbrugh.)*

Lady Brasenose, a London salonnière.

Abraham de Moivre (1667–1754), French-born and French-educated mathematician, spent his adult life from 1687 in England. Fellow of the Royal Society in 1697. Member of anonymous Royal Society Commission of 1712.

Gottfried Wilhelm Leibniz (1646–1716), Leibzig-born, one of Germany's greatest polymaths. Promoted scientific academies including the Brandenburg Society of Sciences ("Berlin Academy"), in 1700 appointed its life president. Trained in law and philosophy, self-taught mathematical genius, eventually invented (independently of, though later than Newton) and published first (prior to Newton) the calculus with notations used to this day, also interested in a mechanical calculating machine. In 1710 published *Théodicée,* rationalizing the existence of evil in a world created by a good God. Universal letter writer (in French, German, and Latin) with more than 1100 correspondents. Mostly in service of the court of Hanover, he never held formal academic teaching positions. Elected Fellow of the Royal Society in 1673 and to the French Academy of Sciences in 1701. Died in Hanover in 1716. *(To be played by same actor as Colley Cibber.)*

Sir Isaac Newton (1642–1727), England's greatest mathematician and natural philosopher, also immersed for decades in alchemy and heretical theology. Enunciated the laws of motion and gravitation and their application to celestial mechanics. Made fundamental contributions to light and color as well as inventing a form of the calculus (termed by him "Method of Fluxions"). Author of two of the most important books in science: the *Philosophiae naturalis principia mathematica (Principia)* and *Opticks.* Fellow of the Royal Society in 1672, President of the Royal Society from 1703 to 1727, in 1669 elected Lucasian Professor of Mathematics at Cambridge University. Appointed Master of the Royal Mint in 1699, and knighted in 1705 by Queen Anne. Notorious for ferocious struggles with scientists (e.g., Robert Hooke and John Flamsteed), but none fiercer and longer than the one with Leibniz. Buried in Westminster Abbey, where his monument was unveiled in 1731. *(To be played by same actor as Sir John Vanbrugh.)*

CURTAIN IMAGE

(The following text should be projected on the curtain or other suitable surface prior to the start of the play.)

An Account of the book entitled
"commercium epistolicum collinii & aliorum"

Published by order of the Royal Society, in relation to the dispute between Mr. Leibniz and Mr. Newton, about the right of invention of the method of fluxions, by some called the differential calculus.

This commercium is composed of several ancient letters and papers. And since neither Mr. Newton nor Mr. Leibniz (the only persons alive who know and remember anything of what had passed in these matters forty years ago) could be witnesses, the Royal Society therefore appointed a numerous committee of gentlemen of several nations to search old letters and papers, and report their opinions upon what they found.

And by these letters and papers it appeared to them that Mr. Newton had the Method in or before the year 1669, and it did not appear to them that Mr. Leibniz had it before the year 1677.

London, 24 April 1712

Act 1, Scene 1

London, 1725. COLLEY CIBBER and SIR JOHN VANBRUGH sit in two comfortable club chairs.

CIBBER: Revenge... *and* scandal? Who can resist such temptation? *(Pause)* Still, I must confess Sir John... your invitation took me by surprise.

VANBRUGH: Why not just "John?" That's what you used to call me.

CIBBER: *(Laughs)* And you used to call me "Colley."

VANBRUGH: So Colley... what's the difference... other than a quarter of a century?

CIBBER: Then you wrote a play that still graces our stage from time to time.

VANBRUGH: *The Relapse. (Pause)* It would never have been written had I not seen the year before the public's lust to see your *Love's Last Shift.*

CIBBER: *(Laughs)* Of course, more than one of our precious critics condemned its... *(Jokingly draws quotes in the air)* "blatantly fleshy treatment of sex."

VANBRUGH: "Giants in wickedness" they called us. *(Dismissively)* Those pygmies of piety. *(Angrily)* And accusing me of "debauching the stage beyond the looseness of all former times." It still rankles.

CIBBER: But revenging yourself after all those years?

VANBRUGH: Some insults continue to fester... and to ooze pus.

CIBBER: You require revenge to lance your boil?

VANBRUGH: It's an efficient method...

CIBBER: From an architect of plays you have become an architect of palaces.

VANBRUGH: A sin?

CIBBER: Not at all! But their scale! First, Castle Howard, then Blenheim—

VANBRUGH: Blenheim Palace demanded it. A fitting tribute to the Duke of Marlborough's victory.

CIBBER: Indeed, indeed... the biggest palace ever built... and garnered you your knighthood. But after all those years, forget... if not forgive... the insults. Why another play?

VANBRUGH: I started as a playwright... I was insulted as a playwright... I wish to end as a playwright... and revenge myself as one.

CIBBER: Through a scandal without sex?

VANBRUGH: Without sex!

CIBBER: Perhaps a dalliance or two?

VANBRUGH: No dalliances!

CIBBER: So how can it be scandalous?

VANBRUGH: Must sex and scandal always be coupled?

CIBBER: It helps... especially on stage.

VANBRUGH: My revenge is aimed at those who want to cleanse our theatre in their holier-than-thou image. Attempting to destroy my reputation... and yours as well!... Aiming to drive our plays from the English stage!... Frothing with indignation in their tracts and pamphlets!... *(Sardonically)* Anointing themselves a "Society for the Restoration of Manners"!... I'll teach them manners! I'll show that real scandal is of the mind.

CIBBER: That will require both artfulness and ambiguity.

VANBRUGH: Hardly the latter. *(Pause)* I have invited you to seek advice... and also your assistance. You are a playwright... an actor... you know how to run a theatre... you are a man familiar with society... someday, you may even become poet laureate....

CIBBER: Enough! You flatter me...

VANBRUGH: You've never held it against me to have built my play *The Relapse* on your success.

CIBBER: The theatre is large enough for both of us.

VANBRUGH: Well put, Colley... and thus a further argument for my proposal. Collaboration by presumed competitors has its merits... a lesson I shall teach through revenge.

CIBBER: But revenge on stage must also divert through a worthy plot. It will require thought.

VANBRUGH: You misunderstood. The plot already exists... in real life. The play and all its scenes.

CIBBER: A drama documenting facts?

VANBRUGH: What about Henry VI? Or King John—one of your favorites?

CIBBER: Will you allow yourself the same liberties as Shakespeare? Taking liberties with facts converts facts into plays.

VANBRUGH: No liberties... just facts in this play.

CIBBER: Take heed, John. I almost see the critics' sneers. *(Pause)* But your play's theme? A theme-less plot would hardly do.

VANBRUGH: Its theme is corruption among the mighty... or call it moral turpitude.

CIBBER: *(Disappointed)* Hardly a novel theme. Again, take Shakespeare—

VANBRUGH: My dear Colley! I am referring to the mighty of the mind... not of the realm. Kings of ideas... not rulers of nations.

CIBBER: And the end of your play? The lesson to be learned?

VANBRUGH: It may well be a play without an end. Sex always ends... but moral deviation? Were we not accused of the same? So let us show them real deviation.

CIBBER: First no sex... and now no end?

VANBRUGH: But there is a scandal!

CIBBER: That may help... but does a scandal always make a play? We thought so then...

VANBRUGH: And will prove so now. The scandal I have chosen is real... indeed all too real. There is no room for ambiguity.

CIBBER: Are its protagonists still alive?

VANBRUGH: All of them!

CIBBER: That warrants care.

VANBRUGH: As well as subtlety... and subtlety takes time—a precious commodity... especially at my age. I'm sixty-one, Colley! Many consider me old.

CIBBER: Nonsense, John. *(Grins)* Though I must admit I was surprised... six years ago... to learn that you had suddenly decided... in your maturity... upon an exploration of marital bliss—

VANBRUGH: How old were you when you succumbed to that temptation?

CIBBER: Promise not to tell. *(Simulates whisper)* Not yet twenty-two!

VANBRUGH: *(Shocked)* How rash!

CIBBER: It was an act of love... but also of madness, bearing in mind that I lacked an income hardly sufficient for one.

VANBRUGH: Perhaps I'm more cautious. I was fifty-five when I proposed to Henrietta.

CIBBER: Lady Henrietta is a handsome woman.

VANBRUGH: In form as well as in figure. And young. *(Pause)* Though not as young as yours.

CIBBER: A wise decision on your part.

VANBRUGH: How so?

CIBBER: My young Catherine was overburdened by fertility. For every child she bore, I had to write a play to support it.

VANBRUGH: Good God! Have you not written at least a dozen plays?

CIBBER: Twenty-five... to be precise... though not all worth remembering.

VANBRUGH: *(Startled)* Your wife bore twenty-five children?

CIBBER: *(Laughing)* Only eleven... but these in such rapid succession that I decided upon withdrawal. *(Pause)* But enough of me... and of my plays. We meet here to talk of *yours*. You are a celebrated playwright. Audiences will recognize your voice.

VANBRUGH: I will conceal my voice by merging it with yours.

CIBBER: My time is at your service, John. *(Pause)* When shall we start?

VANBRUGH: Now.

CIBBER: This moment?

VANBRUGH: I have your attention... so why not hold it by telling you the tale... at least as much of it as I've uncovered so far?

END OF SCENE 1

Act 1, Scene 2

London. 1712. Reception room in Arbuthnot home. CIBBER and VANBRUGH sit in dim background observing the scene.

MRS. ARBUTHNOT: *(Throughout with Scottish accent)* Mr. Bonet, what a pleasure to finally make your personal acquaintance.

BONET: *(Throughout with French accent)* You are most gracious... and so is your husband... for receiving me on such short notice.

MRS. ARBUTHNOT: I must start with an apology... on behalf of Dr. Arbuthnot. An unexpected summons to the Royal Society intervened this morning. He asked that I entertain you until he returns from his meeting with the President. He promised speed.

BONE: *(Disappointed)* Of course, one does not keep Sir Isaac waiting....

MRS. ARBUTHNOT: Quite so. As Master of the Royal Mint he now values only gold more than his time.

BONET: I was thinking of his concern with a new Committee of the Royal Society... a matter unlikely to be familiar to you....

MRS. ARBUTHNOT: On the contrary.

BONET: *(Taken aback)* A Committee convened to adjudicate a delicate matter—

MRS. ARBUTHNOT: Delicacy is a subjective notion... my husband is fond of saying. What is delicate to one may be tedious to another. But since he's not just a physician and savant... but also a writer on human foibles... I take such remarks to heart.

BONET: A wise decision… to accept your husband's perspicuity.

MRS. ARBUTHNOT: I said "take it to heart"… I did not say I always accept it.

BONET: Why question his comment on delicacy… it seems quite unexceptional?

MRS. ARBUTHNOT: But perspicuity is also subjective… even my husband's… as well as yours. You called your Committee's purpose "delicate"—

BONET: *(Forceful)* I consider it exceedingly delicate.

MRS. ARBUTHNOT: I consider it ill advised… if not worse. It demeans two giants attempting in such spiteful fashion to cut each other down to size… not to speak what it does to the rest of you.

BONET: *(Astonished)* Your husband discussed with you our Committee's brief? A matter not even disclosed to all Fellows of the Royal Society?

MRS. ARBUTHNOT: I would venture to go further. Has it even been shared with all members of the Committee?

BONET: I believe so.

MRS. ARBUTHNOT: How can you be so certain… having been appointed but a few days ago?

BONET: I trust you do not take this question amiss: but why would your husband discuss with you delicate… *(Catches himself)* or… if you please… confidential concerns of our Committee?

MRS. ARBUTHNOT: Because I am his wife!

BONET: Yes… but—

MRS. ARBUTHNOT: You do not take your wife into your confidence?

BONET: I have no wife… yet.

MRS. ARBUTHNOT: But if you had one?

BONET: I would not talk about such matters.

MRS. ARBUTHNOT: Yet with my husband... a near stranger... you are prepared to exchange questions that you would keep from your wife? Why? Because you trust my husband?

BONET: I have no reason not to trust him.

MRS. ARBUTHNOT: Yet you'd distrust a wife? *(Pause)* Since you have none yet, I would advise you to choose prudently... as did my husband. *(Pause)* But I'm being carried away. I should have offered some refreshment... rather than contentious thrusts and counterthrusts. May I make up for it now?

BONET: I would prefer to continue with our conversation... though perhaps in a slightly different direction.

MRS. ARBUTHNOT: I would be pleased to oblige... but on what subject?

BONET: Since your husband has been so open with you about our Committee... may I be the same... with a question that I had intended solely for his ears?

MRS. ARBUTHNOT: Of course.

BONET: You intimated that not all members of the Committee are equally well informed about its purpose. You know this for a fact?

MRS. ARBUTHNOT: I do not... and yet feel quite certain of my conclusion. The point at issue solely involves Sir Isaac... or better said, his mathematical work... yet another made the accusation... not even a member of your Committee.

BONET: You are referring to John Keill, whose accusation against Mr. Leibniz prompted the appointment of the Committee?

MRS. ARBUTHNOT: Indeed. But quite difficult mathematics... fluxions and calculus... is at stake here... and especially the question who invented what first. Yet nearly half the members of your Committee are not even mathematicians: Abraham Hill... William Burnet... the Earl of Radnor... Francis Aston... *(Pause)*... and you.

BONET: What about your husband? He is Queen Anne's physician...

MRS. ARBUTHNOT: The best she ever had.

BONET: He is prominent in literary circles—

MRS. ARBUTHNOT: And now collaborating with John Gay and Alexander Pope in a play...

BONET: That I did not know.

MRS. ARBUTHNOT: It's called "Three Hours after Marriage."

BONET: An ambiguous title.

MRS. ARBUTHNOT: Depending on one's view. It's meant to be a comedy.

BONET: *(Astonished)* Physician, man of letters, and now playwright of comedies?

MRS. ARBUTHNOT: *One* comedy... and *(Aside)* his last, I pray.

BONET: But is mathematics another of his talents?

MRS. ARBUTHNOT: According to my husband, very few things are not capable of being reduced to a Mathematical Reasoning which he uses to study probability. Thus, not too long ago, he presented a paper to the Royal Society on the slight excess of male over female births... another of his studies on probability.

BONET: For which he offered an explanation?

MRS. ARBUTHNOT: He concluded that polygamy is contrary to the laws of nature and justice.

BONET: *(Taken aback)* How does that explain the excess of male births?

MRS. ARBUTHNOT: That I never understood. But as I condemn polygamy, I chose not to question him further since his explanation—though perhaps not applicable—suits me. *(Disingenuously)* Was this the question you came to ask him?

BONET: *(Primly)* My question was whether your husband has as yet received the evidence we're asked to weigh. The evidence behind Mr. Keill's accusation. So far, I've received nothing.

MRS. ARBUTHNOT: Forgive my forwardness, Mr. Bonet, but I believe that when you were asked to join the Committee... all of you... you were expected to assent... not weigh—

BONET: *(Disturbed)* How could we assent so quickly... without first weighing?

MRS. ARBUTHNOT: A fair question....

BONET: Is all this inference or knowledge?

MRS. ARBUTHNOT: Inference based on information from an irreproachable source.

BONET: *(Angrily)* In other words, Newton consulted with your husband, but not with the rest of the Committee?

MRS. ARBUTHNOT: He may have spoken to others... most likely to Mr. Moivre.

BONET: *(Angrily)* Who was appointed last!

MRS. ARBUTHNOT: Time of appointment may not relate to rank in the Committee.

BONET: I fail to comprehend—

MRS. ARBUTHNOT: *(Interrupts)* I suspect that comprehension is unnecessary. Tomorrow, your Committee will search for light... an endeavor in which Sir Isaac is preeminent. But your deliberations will focus on the moon reflecting light from the sun. I wonder how many of you will notice that all heat is missing. *(Pause)* In other words... only acceptance counts.

END OF SCENE 2

Act 1, Scene 3

London. 1712. An hour later. Same as Scene 2. CIBBER and VANBRUGH sit in dim background observing the scene. For seamless transition from preceding scene, ARBUTHNOT enters from opposite side to BONET'S exit. Both DR. and MRS. ARBUTHNOT speak with Scottish accents, with hers the stronger.

ARBUTHNOT: Why shouldn't I? *(Pause, then with increasing intensity)* Why? Why? Why? *(Longer pause)* Margaret! Why don't you answer me?

MRS. ARBUTHNOT: I can't.

ARBUTHNOT: But why?

MRS. ARBUTHNOT: I'm afraid.

ARBUTHNOT: Of *me*... your husband?

MRS. ARBUTHNOT: Not *of* you... but *for* you—my husband and the father of my children. I'm afraid of the consequences.

ARBUTHNOT: Nonsense! I'm a Fellow—

MRS. ARBUTHNOT: But he's the President.

ARBUTHNOT: He'll understand when I explain—

MRS. ARBUTHNOT: He may understand—

ARBUTHNOT: You see?

MRS. ARBUTHNOT: But he will never forgive you.

ARBUTHNOT: Nonsense!

MRS. ARBUTHNOT: John... you're being foolhardy.

ARBUTHNOT: An untruth is best contradicted by truth... not another untruth.

MRS. ARBUTHNOT: That will never work with him. Diplomacy? Perhaps. But honesty?

ARBUTHNOT: An unwarranted conclusion!

MRS. ARBUTHNOT: I'd call it logical.

ARBUTHNOT: I see. And the source of your logic?

MRS. ARBUTHNOT: Experience. *(Pause)* And a wife's superior memory.

ARBUTHNOT: *A* wife? Or do you mean *this* wife?

MRS. ARBUTHNOT: You have only one wife... or so I trust.

ARBUTHNOT: You are clever, Margaret—

MRS. ARBUTHNOT: I shan't disagree.

ARBUTHNOT: But at times too clever... for instance this time.

MRS. ARBUTHNOT: Not this time, John. It takes no cleverness—only good memory—to know that Sir Isaac will never accept an honest explanation—however diplomatically delivered—that criticizes him. My words are based on wifely concern.

ARBUTHNOT: I shall not criticize him.

MRS. ARBUTHNOT: Yet you will explain why you cannot sign?

ARBUTHNOT: That I intend to do.

MRS. ARBUTHNOT: He will consider your refusal public criticism.

ARBUTHNOT: How so? I shall explain in private.

MRS. ARBUTHNOT: Your name's absence from that document alone will be sufficient insult.

ARBUTHNOT: I shall prove you wrong.

MRS. ARBUTHNOT: Please, John! Please sign. You cannot afford the risk. He will spit on you—

ARBUTHNOT: Margaret!

MRS. ARBUTHNOT: And then convince you it's raining.

ARBUTHNOT: Nonsense.

MRS. ARBUTHNOT: Have you forgotten his cruelty? As Master of the Mint, Newton applauds the flaying and hanging of many a man who crosses his path.

ARBUTHNOT: The Master of the Mint's duty is to ensure the soundness and safety of our country's coinage. Forgers must be punished!

MRS. ARBUTHNOT: But attend in person the execution of every forger and clipper... and do so for years? Hardly a requirement for an occupant of so high an office.

ARBUTHNOT: *(Attempts humor, which falls flat)* Execution within the Royal Society would be less bloody...

MRS. ARBUTHNOT: ... but surely more painful. *(Disgusted)* Every time I think of him I am suffused by a feeling of formication. *(Mimes gesture as if ants were crawling up her arms)*

ARBUTHNOT: *(Shocked)* Good God, Margaret! Fornication? *(Pause)* With him?

MRS. ARBUTHNOT: *(Primly)* I said *formication*! Have you forgotten your Latin, John? *Formica*—the Latin word for ant.

ARBUTHNOT: And *fornix*... the brothel! Who has taught Latin to whom? Surely not you to me? But what do you mean by formication?

MRS. ARBUTHNOT: The feeling of ants crawling over your skin. In other words... disgusting!

ARBUTHNOT: I am relieved.

MRS. ARBUTHNOT: In that case I shall change. We are expected at Lady Brasenose's.

END OF SCENE 3

Act 1, Scene 4

London. 1712. Same day. Lady Brasenose's salon. CIBBER and VANBRUGH sit in dim background observing the scene.

LADY BRASENOSE: Why? *(Pause, then with increasing intensity)* Why? Why? Why? *(Longer pause)* Mr. Bonet! Why? *(BONET walks to the window, remains quiet, whereupon LADY BRASENOSE assumes formal tone)* Mr. Bonet, did you hear me?

BONET: Yes.

LADY BRASENOSE: Then why do you not reply?

BONET: Because your Ladyship wouldn't understand.

LADY BRASENOSE: I lack intelligence?

BONET: My dear Lady Brasenose... those are your words... not mine.

LADY BRASENOSE: *(Falsetto)* "My dear Lady Brasenose. Not solely your beauty and breeding... it's your brain that always lures me here." *(Resorts to ordinary tone)* Those were your words.

BONET: A long time ago.

LADY BRASENOSE: Is six years that long ago? Long enough for my beauty to have wilted? Long enough for my breeding to have deteriorated? Long enough for my brains to have desiccated? Is that it?

BONET: *(Tired voice)* Please!

LADY BRASENOSE: Then why would I not understand your reason for accepting appointment to the Committee?

BONET: *(Facing her, firmer)* Because you are not a Fellow.

LADY BRASENOSE: I suspect Dr. Arbuthnot will refuse—

BONET: That would be a mistake.

LADY BRASENOSE: He is a Fellow—

BONET: And a fool—

LADY BRASENOSE: *(Sharply)* The doctor is no fool... nor is his wife!

BONET: *(Backs down)* I meant... if what you say is true.

LADY BRASENOSE: Did not Newton himself select him for the Committee?

BONET: Sir Isaac may yet regret it.

LADY BRASENOSE: Is that what counts? Newton's regrets... or lack thereof?

BONET: Yes... that is important. But you would not understand.

LADY BRASENOSE: Because I'm not a Fellow?

BONET: That... and because you do not wish to understand.

LADY BRASENOSE: I cannot be a Fellow. The Royal Society does not believe it needs women—

BONET: Of course we need women... that's why I accepted all your invitations.

LADY BRASENOSE: That had nothing to do with the Royal Society. You only became a Fellow a few months ago.

BONET: Being part of your circle surely helped me become one. Everyone seeks your invitations.

LADY BRASENOSE: Spare me the praise! Instead, let us consider Newton. He's sixty-nine and single... and not just a Fellow, but also the President. I know of no women in his life... not one.

BONET: But that is true of many other men. I'm not married.

LADY BRASENOSE: You are almost thirty years younger. You will marry some day... and of course produce children.

BONET: Why "of course"? Your ladyship is married... yet you have no children.

LADY BRASENOSE: Men produce children... women bear them. *(Pause)* I chose not to bear that load.

BONET: Few women have that choice.

LADY BRASENOSE: Because most men won't grant them that privilege. But I am privileged—

BONET: In more ways than one, my lady.

LADY BRASENOSE: As we all know. Now... you are cautious... but you will marry. You do not dislike women.

BONET: But Sir Isaac does?

LADY BRASENOSE: Even worse... he fears them. He will never marry.

BONET: He took you into his confidence?

LADY BRASENOSE: I do not know of anyone who shares his confidence.

BONET: In that case...

LADY BRASENOSE: But others confide in me.

BONET: That I can believe. I've often heard your salon called London's confessional.

LADY BRASENOSE: I would hardly consider that a tribute... coming from so Protestant a mouth as yours.

BONET: My Lady, accept it as praise, since I—a confirmed Protestant—have visited you so often out of my own free will.

LADY BRASENOSE: Enough of your compliments. My source is a visitor to my confessional, who knows why Newton has been single and will remain so.

BONET: *(Intensely curious)* That I find intriguing! Could you divulge your source's identity?

LADY BRASENOSE: If I did, my salon would turn into a confessional of ill repute if confidentiality is not honored. But there is a matter that should concern you and your Committee. *(Pause)* My dear Bonet. You are the King of Prussia's representative to our court. But how well do you... a diplomat... know Sir Isaac?

BONET: *(Hesitates)* Not well.

LADY BRASENOSE: But you have met him?

BONET: But once.

LADY BRASENOSE: In private?

BONET: At the Royal Society when I signed the book as a new Fellow.

LADY BRASENOSE: I conclude that you do not know him at all! But as a newly appointed Committee member you should be aware of his qualities—

BONET: *(Interrupts)* His merits are well known—

LADY BRASENOSE: *(Interrupts in turn)* As well as his foibles, quirks... and more. For instance, take his fondness for anagrams.

BONET: Anagrams are no rarity in salons. Have we not indulged in them ourselves... right here?

LADY BRASENOSE: I have heard it said that when Newton first thought of his method of fluxions, he wrote it down in his notebook—

BONET: But surely that is not unusual. Where else should he have written it?

LADY BRASENOSE: But disguise it in secret anagrams? Or have anagrams now become the mode in scholarly writings? *(Waves her hand in dismissal)* No matter. But Newton has gone beyond mathematics in that regard. He once showed my source the words *Jeova sanctus unus.* Of course, he would deny it now.

BONET: Why deny it? Surely the Latin words for "God's holy one" are not sacrilegious?

LADY BRASENOSE: Unless you were enamored by anagrams.

BONET: You are now speaking in riddles.

LADY BRASENOSE: Not uncommon for persons skilled in anagrams and knowing that in Latin the letters J and I are used interchangeably.

BONET: *(Annoyed)* What is the answer?

LADY BRASENOSE: "Isaacus Neutonus."

BONET: Sir Isaac's Latin names?

LADY BRASENOSE: Indeed. And should the Master of the Royal Mint... or even the President of the Royal Society... consider himself "God's holy one"? Because he was born on Christmas Day with no father alive? Daubing himself the divine messenger possessed with the confidence of a holy son to construct a picture of God's design for nature?

BONET: But no one means all he says!

LADY BRASENOSE: That may be true of diplomats... like you. But those who know him, will tell you that Newton says all he means. *(Pause)* So why did you do it? Why did you allow Newton... whom you barely know... to persuade you? Why did you sink so low? Why did you—

BONET: *(Interrupts angrily)* I refuse to be questioned in this fashion! Even by you, Lady Brasenose. In time you shall learn the answer. *(Pause)* If not from me, then surely from some one less restrained.

LADY BRASENOSE: A temptation I shall not resist... even if it requires loosening tighter lips than yours.

END OF SCENE 4

Act 1, Scene 5

London. 1725. One week later. CIBBER and VANBRUGH sit in same chairs as in Scene 1.

VANBRUGH: Well, Colley? You had a week to ponder. What do you think?

CIBBER: It's promising… so far.

VANBRUGH: I hear a "but" lurking about.

CIBBER: But where is Newton? I gather it's a disclosure of *his* scandal you wish to forge into revenge.

VANBRUGH: Everyone knows why Newton became President of the Royal Society… the greatest natural philosopher and mathematician of our time. Of course, some asked why he left Cambridge to accept his Majesty's appointment as Master of the Mint.

CIBBER: That is obvious: a great deal of money—

VANBRUGH: Ah, yes. "The love of money is the root of all evil." 1 Timothy 6.10.

CIBBER: Quoting the New Testament is hardly scandalous. Excessive love of money might be… but then we should use it! For instance, his South Sea Company speculation—

VANBRUGH: *(Brusquely)* I prefer not to raise that painful subject. It showed we did not learn from the Dutch tulip mania.

CIBBER: You also bought shares?

VANBRUGH: And lost them all! But Sir Isaac? He had made a 100% profit on his investment as the stock rose. Further evidence of his genius with numbers… hardly a point worth emphasizing in the play.

CIBBER: Like all theatre managers, I search for money... and in that process learn of speculations—

VANBRUGH: With theatre the greatest of them all?

CIBBER: No theatrical disaster ever approached the bursting of the South Sea Bubble.

VANBRUGH: How true. But Newton?

CIBBER: The stock kept rising—

VANBRUGH: And rising... a familiar story even today.

CIBBER: Until even the great Newton... by then Master of the Mint... speculated again.

VANBRUGH: *(Taken aback)* And lost?

CIBBER: All £20,000 of it!

VANBRUGH: That I didn't know.

CIBBER: Shall we use it?

VANBRUGH: *(Wags his head in doubt)* It's tempting... yet all too common... especially today. It will dilute the point I wish to make.

CIBBER: You could relate it to his interest in alchemy. Few know about that.

VANBRUGH: Newton wasn't just interested in alchemy... he was obsessed by it. But he was after the philosopher's stone... the unity of nature... not after gold. Yet Sir Isaac has never written or spoken in public on the subject. Furthermore, it's not relevant to the dispute at issue, which deals with mathematics.

CIBBER: *(Annoyed and impatient)* Then raise the issue... and explain the mathematics! And do it with the protagonists... not the minions circling around him... like moths attracted to the candle's light.

VANBRUGH: Who all get burned! Precisely the approach I wish to take. We have eleven minions... or call them moths if you wish... tainted by this scandal... and all of them Fellows of the Royal Society.

CIBBER: You cannot have all eleven in your play! We cannot afford the expense.

VANBRUGH: I shall only use three.

CIBBER: Were all eleven chosen by Newton?

VANBRUGH: Not formally so... a point we shall make in the play.

CIBBER: Did the Fellows know he had scrutinized the list?

VANBRUGH: Some may be knaves, but none were fools. Brook Taylor, for instance. He was elected to Fellowship on the very day the Committee was established. And who proposed him?

CIBBER: Newton?

VANBRUGH: Of course not! That would have been too obvious. It was John Keill... often called Newton's war-horse... and soon thereafter to become Professor of Astronomy at Oxford... thanks to Newton's aid.

CIBBER: If you need astronomers, why not Edmond Halley?

VANBRUGH: It was Halley who convinced Newton to publish his *Principia Mathematica*. *(Pause)* And one of the first members to join the Committee.

CIBBER: I've met Halley. He's witty... even flippant... qualities that seem to be sorely lacking among your choices—

VANBRUGH: Not true of Arbuthnot! It's been said of him... and I believe justly so... that he has more wit than most... and more humanity than wit.

CIBBER: But not a ladies' man... a flair useful in a play. Edmond Halley... dare I say?... was such a man.

VANBRUGH: And your evidence?

CIBBER: Once on a long sea voyage, a husband and his amply endowed wife, who had so far been unable to conceive, accompanied him. Halley addressed that issue forthrightly and not long after the three had returned to London, the woman had borne fruit. *(Smiles)* Would you call that gossip or proof?

VANBRUGH: *(Laughingly)* Pure conjecture as to cause and effect.

CIBBER: It would work well in a play, but... *(Raises his hand to stop any interruption)* I know...I know: no sex in this one. *(Pause)* Yet you could suggest that sex is akin to mathematics.

VANBRUGH: *(Sarcastic)* I must admit that such resemblance has escaped me... so far. I know of your competence in one endeavor... but both?

CIBBER: If competence in mathematics is required for a playwright, no plays will ever be written about mathematicians.

VANBRUGH: *(Amused)* In that case, enlighten me about the kinship between sex and mathematics.

CIBBER: Both can lead to practical results... even unexpected ones... but that is not foremost in the practitioners' minds when they indulge in it. Most often it is pleasure.

VANBRUGH: Not curiosity?

CIBBER: Satisfying one's curiosity often leads to pleasure. *(Sighs)* But why bring in the King of Prussia's Minister in England... hardly a scandalous occupation?

VANBRUGH: None of my characters have scandalous occupations... least of all Bonet.

CIBBER: Germans are never scandalous. Learned? Yes... Hard-working? Always... Dull? Often... Cruel? Perhaps... But scandalous?

VANBRUGH: Our Bonet is not German. Some would call him Swiss—

CIBBER: Swiss? Good heavens, John! Even worse than German! For the Swiss... what is not prohibited is proscribed. I advise eliminating him from the play.

VANBRUGH: This Bonet is from Geneva.

CIBBER: That is a mitigating fact... possibly even promising. French scandals are the best... and Geneva is right at the border.

VANBRUGH: He studied medicine in Leyden at age 15.

CIBBER: I would never mention that in the play!

VANBRUGH: Because Leyden is in Holland?

CIBBER: I am referring to his age. Precocity is never appreciated on the stage.

VANBRUGH: Would you allow me to refer to the fact that in London Bonet first joined the Society for the Propagation of the Gospel and four years later the Society for the Promotion of Christian Knowledge?

CIBBER: I am beginning to dislike the man—

VANBRUGH: But why?

CIBBER: I cannot stomach pious proselytizers. Besides they do not care for the theatre.

VANBRUGH: That may be so... but his religion is relevant to our play, as I will soon demonstrate.

CIBBER: John... I trust you will not take this amiss *(Takes some pages from his pocket)*, but if the scandal deals with Newton and that German Leibniz... they must appear in your play. You cannot rest your case on surrogates! Here... *(Hands over the pages in his hand)* I used the week's reflection to construct a scene I recommend you insert right now before proceeding further.

VANBRUGH: *(Takes the sheets, folds them and puts them in his pocket)* I shall consider it.

CIBBER: No... let us play it now. I'll do Leibniz and you Newton. After all... he's your man. Without Newton, there would be no play.

VANBRUGH: You're an actor... but I am not.

CIBBER: No matter. You read your text... I've learned mine by heart. Let us be both players and audience to judge the scene's merit.

VANBRUGH: *(Looks at CIBBER for a while, then shrugs his shoulders)* I shall humor you.

END OF SCENE 5

Act 1, Scene 6

London. 1725. Same as Scene 1. CIBBER and VANBRUGH are dressed in costumes appropriate to LEIBNIZ and NEWTON, whom they play. VANBRUGH holds play text in his hands and pretends to read NEWTON'S lines. CIBBER uses German accent while playing the role of LEIBNIZ.

LEIBNIZ: So we finally meet tête-à-tête, Mr. Newton. Finally!

NEWTON: There is nothing that I desire to avoid in matters of Philosophy more than contention, nor any kind of contention more than one in print.

LEIBNIZ: Yet the accusation *was* made in print!

NEWTON: I wrote none.

LEIBNIZ: What about Mr. Keill—

NEWTON: A distinguished Fellow of the Royal Society... and soon to become... upon my recommendation... Savilian Professor of Astronomy at Oxford.

LEIBNIZ: Distinguished? Bah! Keill is one of your sycophants.

NEWTON: How dare you?

LEIBNIZ: How dare *he*? Keill insults me like a fishwife by fabricating suspicions that I have won fame not by the straight road but by devious practices. No fair-minded or sensible person will think it right that I, at my age, and with such a full testimony of life, should state an apologetic case for it, appearing like a suitor before a court of law, against Mr. Keill. *(Increasingly louder)* I, Gottfried Wilhelm Leibniz, whose invention contains the application of all reason... a judgment in each

139

controversy... an analysis of all notions... a valuation of probability... a compass for navigating over the ocean of our experiences... an inventory of all things... a table of all thoughts... a general possibility to calculate everything. *(Takes audible deep breath)* When I published the elements of my calculus in 1684, there was assuredly nothing known to me of your discoveries in this area, beyond what you had formerly signified to me by letter.... But as soon as I saw your *Principia*, I perceived that you had gone much further. However, I did not know until recently that you practiced a calculus so similar to my differential calculus. So what is your view, Mr. Newton?

NEWTON: *(Aside, furious whisper)* That viper in my brain... that Leibniz... not content with deriding my invention of the fluxions, now presents himself to the world as inventor of the calculus! *(Louder with faked calm)* I had no hand in beginning this controversy.

LEIBNIZ: Ha!

NEWTON: Mr. Leibniz! In letters exchanged between myself and you ten years ago, I indicated that I possessed a method of determining maxima and minima... and concealed the same method in transposed letters—which, when correctly arranged—express this sentence *(Slow and forceful tone)* "*Given any equation involving fluent quantities, to find the fluxions, and vice-versa.*"

LEIBNIZ: *(Sardonic)* Ha... ha! If all knowledge were transmitted in 70 transposed letters... *(Extremely fast and sarcastic)* five As, two Cs, one D, seven Es, three Fs, one G, nine Is, three Ls... no less than ten Ns!... four Os, two Qs, one R, three Ss, six Ts, four Us, five Vs and then one X and one Y... where, I ask... where would mathematics or natural philosophy... indeed human knowledge stand now? Are anagrams in science honest? Or are they just a joke? *(Pause)* As I find no H... as in "honesty" or "humor"... nor a J... as in "joke" in your anagrammatic alphabet, neither honesty nor humor could have been the motivation. *(Sardonic laughter)* Indeed, as there is no letter M, even mathematics is precluded!

NEWTON: Once more I ask... how dare you?

LEIBNIZ: Did you not write in 1676: "Leibniz's method of obtaining convergent series is certainly extremely elegant, and would sufficiently display the writer's genius even if he should write nothing else." *(Pause)* Well, Mr. Newton?

NEWTON: One my greatest lapses of judgment.

LEIBNIZ: Are you accusing me of poaching... of trespassing... on English turf? Or perhaps stealing?

NEWTON: Call it what you wish! I first bit into this English apple... and expected to eat it at my leisure.

LEIBNIZ: An apple already bitten... especially an English one... does not attract me. I made German applesauce... so that others could taste it! Whatever we call it... fluxions or calculus... it will be the crown jewel of all mathematics.

NEWTON: Indeed it will.

LEIBNIZ: So for once we agree! But how will that jewel be described? Must I remind you that when you finally chose to launch your "method of fluxions" in print, few equated it as my "infinitesmal calculus." Your terminology was a jargon of flowing points and lines... your so-called fluents. And their rate of change... you called "fluxions." Your adding or subtracting dots over letters to represent *(Derisive)* "fluxions of fluxions or fluents of fluents" is the clumsiest of clumsy notations. *(Forcefully)* Mine was algebraical; my language fresh and clear using the words "differential"... "integral"... and "function." I do not find these in your writings!

NEWTON: My question... *(Catches himself)*... I mean Mr. Keill's question, is who discovered the method first. Priority is exclusive. It is an absolute, quantifiable fact.

LEIBNIZ: Quantifiable? Is that not carrying mathematics too far?

NEWTON: One man is first! Be it by years, weeks, hours, or even minutes.

LEIBNIZ: *(Sarcastic)* I see! The Master of the Mint now turning into Master of the Calendar.

(Exits by returning to the chair where CIBBER sat before and removes LEIBNIZ'S coat to turn into CIBBER. Observes last speech of NEWTON from that chair.)

NEWTON: Leibniz will yet rue the day when he issued this challenge. Whether he found the Method of Fluxions... *(Disdainful)* his calculus... by himself or not is *not* the question. The newly appointed Committee of the Royal Society will only deal with the question who was the *first* inventor. And I shall see that they do not stray from that narrow path! *(Pause)* Those who have reputed Mr. Leibniz the first inventor, know little or nothing of his correspondence with Mr. Collins and others long before. I shall order this correspondence to be brought before the Committee, which is numerous and skillful and composed of gentlemen of several nations. On this point, the Committee will treat Leibniz as second inventor, because as I have said before *(Louder)*... and now say again... *(Slow and loud) second inventors have no rights!* None*! (Turns abruptly and walks toward CIBBER, in the process removing NEWTON'S coat)*

CIBBER: You are a born actor! More Newtonian than Newton.

VANBRUGH: *(Pleased)* I thank you for the compliment.

CIBBER: Well?

VANBRUGH: Yes?

CIBBER: Will you use it?

VANBRUGH: We shall see. I'm of mixed mind. But whether we use the scene or not, we shall end the first Act here.

END OF SCENE 6

END OF ACT 1

Act 2, Scene 7

London. 1712. Lady Brasenose's salon with food laden on a side table. By contrast to BONET (and later ARBUTHNOT), MOIVRE is dressed in threadbare clothes. CIBBER and VANBRUGH sit in dim background observing the scene.

MOIVRE: *(Somewhat overwhelmed while looking around)* My first time in Lady Brasenose's salon. *(Bitter)* Of course, an unlikely venue for a mathematician. *(Turns to BONET)* We seem to be the only guests.

BONET: We are early. Deliberately so. I wanted to speak to you... alone.

MOIVRE: *(Moves to side table, pointing to food)* May I?

BONET: *(Reluctantly)* No one will see you.

MOIVRE: *(While hungrily starting to eat)* I suspect you will be the only guest I shall know here.

BONET: You may meet another.

MOIVRE: I owe this invitation to you, I presume?

BONET: Not entirely. Lady Brasenose is a curious lady—

MOIVRE: But mathematics? This would extend her curiosity beyond normal bounds... at least among her social circles.

BONET: I have yet to find limits to Lady Brasenose's lust for knowledge.

MOIVRE: She lusts?

BONET: Vigorous desire, then.

BONET: *(Moving toward a chair and gesturing to MOIVRE, who is still wolfing down food, to join him)* Monsieur Moivre... you know the reason for this meeting?

MOIVRE: *(Quickly takes another bite and then surreptitiously puts some food, perhaps a roll, into his pocket)* The Committee?

BONET: Of course... and yet there's more.

MOIVRE: *(Manages to put another food item into his pocket before reluctantly joining BONET)* It would be an honor to be of service to the Minister of the Prussian King.

BONET: Ah, yes... Prussia. Only some ten years ago, our King, Frederick III, created our own Academy in Berlin. You know its President?

MOIVRE: Leibniz... a great man... one of Europe's greatest savants.

BONET: *(Ironic)* An opinion he's likely to share with you.

MOIVRE: Through his friend... the great Jean Bernoulli... I had hoped that Leibniz would secure a university chair on my behalf. My request fell on deaf ears.

BONET: You know, of course, that his Royal Highness appointed Leibniz President for Life... and with a handsome stipend.

MOIVRE: We have no President for life in the Royal Society... nor does he receive remuneration.

BONET: *(Slightly ironic)* The Master of the Royal Mint can readily afford the unadulterated honor of being President.

MOIVRE: *(Aside)* With gold already in his pockets. *(Louder)* Of course you are a member of your Berlin Academy...

BONET: *(Stiffly)* I am not.

MOIVRE: *(Taken aback)* Yet you are a Fellow of the Royal Society.. *(Pause)* Would it be discourteous to inquire why you are not a member of your own Academy?

BONET: Because of Leibniz... yet I shall become one... and soon I trust... in spite of him.

MOIVRE: In which class? Not mathematics, I presume?

BONET: Theology.

MOIVRE: For which Leibniz will propose you?

BONET: *(Bitter laugh)* More likely oppose me. But may I return to my question? Why was I chosen... some three weeks after Halley, Arbuthnot and the others?

MOIVRE: But not *all* others! Taylor, Aston, and I were only invited two days ago... now making us a Committee of eleven.

BONET: And why eleven Fellows?

MOIVRE: Perhaps precluding a Judas among Newton's Apostles? *(Quick)* Of course, I'm only jesting. *(Pause)* But why ask me?

BONET: You've been a Fellow for some years—

MOIVRE: Fifteen... to be precise—

BONET: Precision befits a mathematician... which I am not. The Committee's charge concerns mathematics... so why choose me who is most deficient in this field? And who did so? I only received a letter from the Secretary without stating a reason.

MOIVRE: *(Coyly)* The reasons may be murky... but they will become clear during our first meeting. If it had been solely for my mathematical competence... which I claim openly... I should have been among the first group... among Edmond Halley... for whom, I wager, a comet will yet be named... or William Jones, who sensibly introduced the symbol π... *(Pause)* Yet I—though a mathematician—was among the very last... even after you.

BONET: You know the reason?

MOIVRE: At age twenty, after incarceration for refusing to convert to Catholicism, I fled to England... yet they still call me French. *(Bitter)* As a Huguenot émigré, I eke out a living from tutoring students, from publications...

(Goes to buffet table to pick another item of food, then quickly eats it)

... even from solving problems of chance in coffee houses or calculating odds for gamblers... but I have yet to find a true patron to open the door to a position of merit... in this country or the Continent. But now... for the first time... the burden has turned into an advantage... that I shall use to the fullest! The President needs foreigners—

BONET: ... but surely also men who are equipped to judge the issue at hand. What about mathematicians from Holland or Italy or France or—

MOIVRE: Switzerland! For instance, Jean Bernoulli... perhaps the most discerning of them all. But for this Committee, our President needs Fellows on his side... that he can trust. Jean Bernoulli... Leibniz's most forceful defender... is not one of them! Fortunately for Newton, Bernoulli is not yet a Fellow.

BONET: And you are a Fellow whom he can trust?

MOIVRE: For me the creator of the *Principia Mathematica* can do no wrong. Let me recount how I discovered it. Calling one day on the Earl of Devonshire, I saw in the antechamber a copy that Newton had come to present to the Earl that very day. I opened the book and found to my astonishment that, strong as I thought myself to be in mathematics, I could only just follow the reasoning. The following day, I procured a copy and tore out the pages.

BONET: A brutal way of attacking an argument.

MOIVRE: In this instance, the supreme homage... one that I had never paid before.

BONET: I wonder if the author would consider it as such.

MOIVRE: Sir Isaac does, since he knows the state of my affairs. I am obliged to work almost from morning to night... teaching my pupils and walking. Since London is very large, much of my time is employed solely in walking. *(Bitterly points to his badly worn shoes or boots)* That is what reduces the profit I can make and cuts into my leisure for study. But by tearing leaf after leaf from the *Principia* and carrying a few at a time in my pocket, I could peruse it on my walks and other intervals. *(Pause)* Soon thereafter, I was elected to Fellowship in the Royal Society.

BONET: *(With touch of irony)* I wonder how many other Fellows were elected for mutilating a book?

MOIVRE: *(Sharply)* Whatever the reason, I was grateful to have been elected.

BONET: You are a tutor. Could you offer a simple definition? I would be loath to admit openly my ignorance of fluxions when the Committee meets.

MOIVRE: Call fluxions the velocities of evanescent increments... of infinitesimals. Is that simple enough? *(He laughs upon seeing BONET'S dubious expression)*

BONET: You mean too small to perceive?

MOIVRE: So tiny... that if a quantity is increased or decreased by an infinitesimal, then that quantity is neither increased nor decreased.

BONET: You mean zero?

MOIVRE: Larger than that... yet smaller than any other number.

BONET: If that is simple...

MOIVRE: Then try this. What Sir Isaac called the method of fluxions, Leibniz termed calculus—a method that finally related time with space.

BONET: A good tutor should be more specific.

MOIVRE: *(Hardly repressing his impatience)* A method determining the rate of change at any moment of a quantity that itself is changing in relation to another quantity, which Leibniz calls a function.

BONET: Still not specific enough.

MOIVRE: *(Under his breath) Mon dieu!* How about the distance covered by a falling object being a function of the time it had been falling?

BONET: Better.

MOIVRE: *(Relieved)* In that case, I congratulate you. You have mastered what Leibniz called the differential calculus.

BONET: Is there another?

MOIVRE: Integral calculus.

BONET: The reverse of differential?

MOIVRE: *(Derisive)* Very good. *(Didactic)* Newton called it the "inverse method," by which he meant that a whole can be reconstructed from a given value at an instant. From the rate of change at a point, one can derive a line. From a line, the area it defines—

BONET: And from the area?

MOIVRE: Enough! Your contribution to the Committee's deliberation will not depend on your understanding further details. A quick tutorial is insufficient. But for a number of years now, the calculus of probability has caught my attention. First, to study gambling odds but then to address the probability of life itself... to calculate annuities and similar properties... even the date of my own death.

BONET: Your own?

MOIVRE: Now, at age forty-five, I have increasing need of sleep... very small increments each night. I shall pass into eternal sleep when the total reaches twenty-four hours... which I calculate will occur at age eighty-seven.

BONET: But your wife may keep you awake from time to time... thus ruining your arithmetical progression... and prolonging your life.

MOIVRE: I am not married.

BONET: Your future wife then. You said... you're barely forty-five. I am forty-two and not yet married, but I am certain to do so before long—

MOIVRE: *(Interrupts almost angrily)* I shall not marry... because I cannot afford a wife.

BONET: *(Compassionately)* Are numbers... and mathematics... your whole life?

MOIVRE: There is religion!

BONET: That I take for granted. I meant other than mathematics <u>and</u> religion.

MOIVRE: l shall never give up my taste for literature. I know many works by heart... especially Rabélais and Molière.

BONET: So you frequent the theatre?

MOIVRE: I cannot afford it. But I could recite for you the entire *Misanthrope—*

BONET: *(Hastily)* Some other time. But now to the matter at hand: to what circumstances do I owe my selection to the committee?

MOIVRE: Are you aware of Leibniz's defense against Keill's accusation?

BONET: I do not trust Mr. Keill. *(Quick)* Admittedly, a judgment based on short acquaintance... and a remark lacking in diplomacy—

MOIVRE: *(Interrupts quickly)* A remark I shall claim never to have heard. *(Performs a mocking bow)* Diplomacy is thus restored to the King of Prussia's Minister in London. *(Pause)* You asked why you were invited to join the Committee? You would not be pleased to hear my explanation.

BONET: Still... may I hear it?

MOIVRE: They want a foreigner, who does not understand the issue—

BONET: And thus can be manipulated?

MOIVRE: I fear that is true.

BONET: But you... who understands the issue... are also a foreigner.

MOIVRE: They assumed my poverty would make me pliant.

BONET: And are they right?

MOIVRE: Poverty breeds cunning. But you, Monsieur? Do you not resent being picked solely for your foreign credentials and ignorance of mathematics?

BONET: *(Laughs)* You're not one to mince words! But your frankness deserves a frank answer. Under ordinary circumstances, I would have taken it as an insult. But not this time.

MOIVRE: What is different?

BONET: Diplomats often adjust their agenda to the circumstances facing them.

MOIVRE: You now find membership of the Committee convenient for your purposes?

BONET: Extremely so. *(Sees ARBUTHNOT approaching)* Ah... here comes Dr. Arbuthnot... another guest of your acquaintance.

ARBUTHNOT: *(Greets both men, then addresses BONET)* You were right... Lady Brasenose wants to know—

BONET: Not only know... she also wishes to advise.

ARBUTHNOT: *(Smiling)* According to my wife, Lady Brasenose intends to judge.

BONET: She knows Lady Brasenose?

ARBUTHNOT: *(Still smiling)* All too well.

BONET: It's true... on more than one occasion; I have found her Ladyship's advice and judgment to be indistinguishable.

MOIVRE: You gentlemen seem to know more about this invitation than I. Would you enlighten me?

ARBUTHNOT: That will hardly be necessary. *(Points to approaching LADY BRASENOSE)* Lady Brasenose does not waste time.

LADY BRASENOSE: *(Offers her hand successively to BONET, ARBUTHNOT, and finally MOIVRE)* How kind of you not to think of some excuses—

BONET: My lady! Have I ever done so before?

LADY BRASENOSE: Our last contretemps might have caused you to reconsider.

BONET: It was a minor squall... not thunder and lightning.

LADY BRASENOSE: You are diplomatic... no wonder your King sent you to London... and keeps you here... for our pleasure and, of course, his benefit. *(Turns to ARBUTHNOT)* And you, my good doctor? You could have claimed concern for a patient.

ARBUTHNOT: *(Playful, yet slightly deferential)* Only a royal patient could have been the cause. Fortunately, her Majesty is well. *(Pause)* Besides... my wife thought I would benefit from a different view.

LADY BRASENOSE: Different in what regard?

ARBUTHNOT: Different from her view.

LADY BRASENOSE: Mrs. Arbuthnot is a wise woman. But what is her counsel?

ARBUTHNOT: To yield... in other words to sign.

LADY BRASENOSE: *(Archly)* Well... mine is not! *(Abruptly turns to MOIVRE)* And you, Mr. Moivre? What brings you here?

MOIVRE: *(Startled)* Your ladyship's invitation.

LADY BRASENOSE: *(Dismissively)* Of course! I beg your pardon. *(Brief pause)* You're a skillful mathematician... I am told... but were you not appointed just two days ago to the Committee? Why did he wait so long?

MOIVRE: *He?*

LADY BRASENOSE: *(Ironic)* Would you prefer me to say *"they?"* If so, I shall oblige you... though it may be incorrect. *(Looks him over)* So why did they?

MOIVRE: *(Bitter)* Because they do not consider me English!

LADY BRASENOSE: Undoubtedly also the reason why my diplomatic friend *(Points to BONET)* was chosen... unless *(Smiles coquettishly)* he hides from me a competence in mathematics of which I was hitherto unaware—

BONET: Few things escape Lady Brasenose.

LADY BRASENOSE: True so far... and I hope still for years to come. So why, Mr. Moivre? There are other Fellows who are distinguished mathematicians and yet not English: *(Turns to MOIVRE)* An obvious choice would be Guido Grandi—

MOIVRE: *(Explosively)* Out of the question!

LADY BRASENOSE: Because he is Italian? Is that not foreign enough... or perhaps too foreign?

MOIVRE: In England not being English is already too foreign. But Grandi was the first to introduce Leibniz's calculus to Italy.

LADY BRASENOSE: *Touché...* on both counts. But back to my question: why were you appointed?

MOIVRE: *(Vexed)* I already told your Ladyship—

LADY BRASENOSE: Because you are a mathematician, a Fellow... and not considered British? None of those reasons would have caused him... *(Pretends to catch herself)...* I beg your pardon... I meant *them...* to appoint you but three days ago! Barely in time for tomorrow's gathering of your Committee... the *first...* and likely also *last* meeting!

BONET: *(Irritated)* Lady Brasenose! How could you have knowledge of an event that has not yet occurred?

LADY BRASENOSE: Why not ask Dr. Arbuthnot... or his wife, whom you met yesterday?

BONET: These days, news seems to reach my Lady even faster than usual.

LADY BRASENOSE: You seem to forget that you now live in England—an assembly of voluntary spies. *(Turns to MOIVRE)* So what is the reason for your appointment? Not even Mrs. Arbuthnot could hazard a guess.

MOIVRE: Nor will I.

LADY BRASENOSE: So why did you accept?

MOIVRE: Your Ladyship addresses me like a petitioner... rather than as your guest. Since I am not the former... nor apparently the latter, I take my leave.

(MOIVRE is about to exit when he stops)

But I shall answer your question with one word: "Poverty"... or perhaps two words: "sheer poverty." Your Ladyship's hospitality... or absence thereof... will not resolve it. Membership on the Committee might. *(Exits)*

BONET: My Lady... you pushed too hard.

LADY BRASENOSE: *(Quietly)* And now regret it.

BONET: Would you allow one question?

LADY BRASENOSE: How can I refuse you... after all my questions?

BONET: Why did you invite us?

LADY BRASENOSE: I fear Newton is making your bed and is about to blow out the candle to put your Committee to sleep.

BONET: So what is your aim?

LADY BRASENOSE: Most men can deal with conflict... but the real test is how authority is deployed when all can see it. I know you well enough... both of you... not to have to hide my dislike of Sir Isaac. *(Pause)* I want to keep the candle lit.

ARBUTHNOT: A worthy purpose that my wife would applaud.

LADY BRASENOSE: Yet you said she wanted you to yield.

ARBUTHNOT: She's afraid of Newton's wrath... yet wants her husband to be ashamed... and shame is hidden in the dark.

LADY BRASENOSE: Your wife's concerns are colored by affection... and morality. Mine... by morality and curiosity. All your Committee represents for Newton is a collection of barkless watchdogs. But such dogs expect to be fed. I wonder what Sir Isaac has in mind? Not all canines have the same appetite. For instance, poor Moivre is likely to be thankful for some scraps. *(Gestures toward side table with food)* But what about you, my dear Bonet... and your confrères? *(Long pause)* Will none of you bark... and look elsewhere for sustenance?

END OF SCENE 7

Act 2, Scene 8

London. 1712. Arbuthnot home. MRS. ARBUTHNOT paces impatiently. Looks up as her husband enters. CIBBER and VANBRUGH sit in dim background observing the scene.

MRS. ARBUTHNOT: Well?

ARBUTHNOT: It's done.

MRS. ARBUTHNOT: Who was there?

ARBUTHNOT: All eleven.

MRS. ARBUTHNOT: No one else?

ARBUTHNOT: Newton.

MRS. ARBUTHNOT: Of course... but who else?

ARBUTHNOT: No one.

MRS. ARBUTHNOT: That was clever.

ARBUTHNOT: Newton is clever... but also cautious. Why invite unnecessary witnesses?

MRS. ARBUTHNOT: No other supporters? Not even the kind Bernoulli called "Newton's servile sycophants?"

ARBUTHNOT: The Committee is already inundated with them. Besides, Bernoulli called them "Newton's toadies."

MRS. ARBUTHNOT: Hardly more complimentary... considering that your friend, Mr. Pope, referred to another human toad as "part froth, part venom." But would it not have been more politic to include in the Committee some Fellows less beholden?

ARBUTHNOT: There were a few.

MRS. ARBUTHNOT: Bonet?

ARBUTHNOT: He's one.

MRS. ARBUTHNOT: And you.

ARBUTHNOT: *(Tired nod)* And I.

MRS. ARBUTHNOT: *(Impatient)* Tell me what happened.

ARBUTHNOT: I started out on the wrong foot.

MRS. ARBUTHNOT: You mean with honesty?

ARBUTHNOT: *(Nods)* Does truth not bear the same relation to understanding as music does to the ear or beauty to the eye?

MRS. ARBUTHNOT: Newton is concerned with understanding the universe. That truth concerns him… but no other music reaches his ear. I had warned you, John. *(Reaches over to pat his hand or for some other gesture of affection)* What did you say?

ARBUTHNOT: I quoted Francis Bacon: "There is little friendship in the world… and least of all between equals."

MRS. ARBUTHNOT: And?

ARBUTHNOT: I wanted to ask, "Why not prove Bacon wrong?"… but Newton stopped me.

MRS. ARBUTHNOT: What did he say?

ARBUTHNOT: Nothing.

MRS. ARBUTHNOT: But you said he stopped you.

ARBUTHNOT: He pointed to Hill.

MRS. ARBUTHNOT: Why Abraham Hill?

ARBUTHNOT: He's the oldest… almost eighty.

MRS. ARBUTHNOT: *(Dismissive)* And one of the toadies.

ARBUTHNOT: *(Tired)* Not any more than most of us.

MRS. ARBUTHNOT: And what did the oldest toady say?

ARBUTHNOT: That the Committee's concern was with superiority... not equality... of British science. Friendship was irrelevant. If it was proper for Germans to pin on Leibniz another's garland, it is proper for Britons to restore to Newton what is really his own.

MRS. ARBUTHNOT: He said that... in front of Newton?

ARBUTHNOT: He hardly had to... most understood... though perhaps not Bonet.

MRS. ARBUTHNOT: And that was it? *(Not getting any response she continues more exasperated)* John! I've never had to push you like this. Do you not trust me?

ARBUTHNOT: It's a matter of shame... not trust.

MRS. ARBUTHNOT: *(Warmer)* Then confide in your wife.

ARBUTHNOT: I could not help but think of John Flamsteed—

MRS. ARBUTHNOT: You are no friend of his...

ARBUTHNOT: Nor his enemy.

MRS. ARBUTHNOT: Even after you demanded... on behalf of Newton... that Flamsteed deliver his life's work—his lunar tables—to his bitterest enemy?

ARBUTHNOT: I did that at her Majesty's command—

MRS. ARBUTHNOT: After furious prompting by Newton. He used you, John!

ARBUTHNOT: In your eyes he probably did.

MRS. ARBUTHNOT: Newton hates Flamsteed... in spite of his position as Astronomer Royal.

ARBUTHNOT: *(Nods tiredly)* In spite... and because of it. *(Stronger)* Though hardly a justification to have the Astronomer Royal ejected from Fellowship of the Royal Society for late payment of fees. *(Pause)* But that was not the reason I thought of Flamsteed. He once sent me a note, writing, *"Those that have begun to do ill things, never blush to do worse to secure themselves."* I hoped then that he meant Newton... and now I know he did.

MRS. ARBUTHNOT: Because you were presented with the finished report before your Committee had even met?

ARBUTHNOT: Much worse... much, much worse! Newton's conceit exceeded perversity. *(Long pause)*

MRS. ARBUTHNOT: *(Impatient)* How could it be worse? John! You must tell me!

ARBUTHNOT: Newton alone had written it—

MRS. ARBUTHNOT: *(Shocked)* That I cannot believe! Newton could not have been that brazen.

ARBUTHNOT: He was... and cunningly termed the report *"commercium epistolicum collinii & aliorum"*—in plain English, "An exchange of letters between Collins and others."

MRS. ARBUTHNOT: But Collins is dead!

ARBUTHNOT: Letters written to the *late* John Collins and other deceased correspondents by Leibniz and Newton... and selected by Newton... to bolster his case in his own words without contradiction by the dead.

MRS. ARBUTHNOT: That is barefaced. And you were asked to sign... without further debate?

ARBUTHNOT: All of us were.

MRS. ARBUTHNOT: Thank God you yielded... even though I find it utterly humiliating. But I would not have wanted you to suffer Sir Isaac's wrath. We both know his unparalleled cunning—

ARBUTHNOT: And cruelty.

MRS. ARBUTHNOT: What happens now that you have signed Newton's *Commercium*?

ARBUTHNOT: I didn't say that pen was set to paper!

END OF SCENE 8

Act 2, Scene 9

London. 1712. Lady Brasenose's salon without side table. Two days after Scene 7. LADY BRASENOSE, DR. ARBUTHNOT, and BONET in animated—even contentious—conversation. CIBBER and VANBRUGH sit in dim background observing the scene.

LADY BRASENOSE: Now tell me: did he do it?

ARBUTHNOT: Why ask us?

LADY BRASENOSE: Because you were there!

ARBUTHNOT: But so were nine others.

LADY BRASENOSE: I had more confidence in you... both of you.

BONET: My Lady uses the past tense. You have no confidence now?

LADY BRASENOSE: *(Sharp, yet smiling)* You listen to nuances... is that another reason why your King sent you to London?

BONET: In diplomacy, precision in language leads to imprecision in meaning.

LADY BRASENOSE: *(Laughs outright)* I am tempted to pursue this line of conversation... it does suit a salon. But I shall resist the temptation. *(Turns to ARBUTHNOT)* My dear doctor. I presume you took your wife into your confidence.

ARBUTHNOT: Reluctantly.

LADY BRASENOSE: Reluctance is no virtue in my eyes—

ARBUTHNOT: Nor did I claim it so.

LADY BRASENOSE: So out with it! The report is at the printers. Was your Committee's decision unanimous?

BONET: It was.

LADY BRASENOSE: Without giving notice to Leibniz? Without inviting him to offer documents in his defense?

ARBUTHNOT: Without such actions.

LADY BRASENOSE: Yet you all signed? Shame on you!

BONET: You're overlooking the nuances, Lady Brasenose! I said the decision was unanimous—

ARBUTHNOT: But we did not sign.

LADY BRASENOSE: *(Taken aback)* How did you two accomplish that?

BONET: None signed!

LADY BRASENOSE: Sir Isaac is more malleable than I thought. He never compromises as Master of the Mint.

ARBUTHNOT: Not more malleable... just more subtle as President of the Royal Society. It goes with that office... controlling minds rather than coinage.

LADY BRASENOSE: More subtle or more devious? *(Makes dismissive gesture)* No matter. What caused his change of mind?

BONET: *(Points to ARBUTHNOT)* Our doctor's diplomacy.

LADY BRASENOSE: If our physicians now turn into diplomats, what happens to our diplomats?

ARBUTHNOT: I'm afraid my Lady's question is not applicable to the case at hand. Both the physician *(Points to himself)* and the diplomat *(Points to BONET)* chose prevarication.

BONET: Dr. Arbuthnot... you're too severe.

ARBUTHNOT: *(Turns to BONET)* Am I? What is a prevaricator in your eyes?

BONET: A quibbler... or equivocator.

LADY BRASENOSE: In other words... a diplomat.

ARBUTHNOT: *(Quietly)* We were cowards... but hardly diplomats.

LADY BRASENOSE: Cowardice and diplomacy are not mutually exclusive! But Dr. Arbuthnot! I've never before seen you wear a hair-shirt... at least not in my house. It's time you discarded it. Enlighten me... both of you!

ARBUTHNOT: It was clear to me... already when I met the President alone... that not a word would be changed in the report... and that it would be published even if some of us protested.

LADY BRASENOSE: And you were unwilling to be such a protester?

ARBUTHNOT: I have never had an appetite for disputes. This time, even my wife advised against it. For Sir Isaac, such a protester would be an apostate... in his eyes unforgivable in the Church and in the Royal Society. Think of John Flamsteed or Robert Hooke. He destroyed them. *(Tired voice)* So why even attempt it?

LADY BRASENOSE: For principle's sake?

BONET: Principles exist to be broken... at least at times.

LADY BRASENOSE: The diplomat is speaking.

BONET: Or the churchman. *"Thou shalt not kill"* has never prevented religious wars.

LADY BRASENOSE: *(Impatient)* I want to hear about the prevarication... not its rationalization.

BONET: Dr. Arbuthnot's proposal was adroit. What Newton needed... what Keill and the other lackeys needed—

ARBUTHNOT: *(Mildly remonstrating)* At the meeting I said "Fellows"... not "lackeys." Using that word would not have served my purpose.

BONET: I now allow myself that liberty, since we're *entre nous*. But to proceed: what was needed was a published unanimous report... *(Pause)* but for that the identity of the Committee could remain undisclosed. *(Assumes precious, ironic tone)* A *"Numerous Committee of Gentlemen of Several Nations"* was surely adequate—

LADY BRASENOSE: Using Sir Isaac's own words?

BONET: Precisely. And once granted that, how could the President deny the logic of Dr. Arbuthnot's request that unanimity by vote of an anonymous Committee surely need not be confirmed by signature? Publication is enough.

LADY BRASENOSE: Like the death warrant for Charles I?

ARBUTHNOT: An apt comparison. Killing a scholar's reputation is also murder.

BONET: Be that as it may. *(Pause)* Dr. Arbuthnot's proposal was carried unanimously.

LADY BRASENOSE: *(Admiringly)* I never thought prevarication could be so effectual. I stand enlightened. *(Turns to BONET)* But you could have been the honorable exception to unanimity. You have the least to fear of Newton. His wrath will not follow you to Geneva.

ARBUTHNOT: Lady Brasenose... and Mr. Bonet. Forgive me, but I must take my leave. A patient waits who must not be kept waiting.

(Exits.)

LADY BRASENOSE: *(Turning to BONET)* Now that we are alone, I trust you will answer honestly.

BONET: I did not vote *for* Newton... I voted *against* Leibniz.

LADY BRASENOSE: You judged Leibniz a plagiarizer?

BONET: I am not qualified to pass judgment in mathematics.

LADY BRASENOSE: But that was the issue!

BONET: To me it was a question of higher truth... not subject to adjustments, however infinitesimal. In a religious calculus, adjustments cannot be tolerated. Leibniz's latest writings justify theodicy, which I find unacceptable. *(Pause)* And so does Newton.

LADY BRASENOSE: *The Odyssey*? How does Homer enter the argument?

BONET: *(Impatiently)* Of course not Homer. *(Spells it, loud and slowly)* T-H-E-O-D-I-C-Y.

LADY BRASENOSE: *(Laughs)* Theodicy! My dear Bonet, I have often complimented you on your English, but once in a while your accent does mislead. As in theodicy. But you share Newton's views on religion?

BONET: *(Quick)* We both abhor reunification of Protestantism and Popery. Yet Leibniz, though claiming to be Lutheran, moves easily in Catholic circles... and wishes more of us to do so. *(Vehement)* That crypto-Catholic! And on theodicy, Newton and I see eye to eye.

LADY BRASENOSE: On theodicy, I'd take Leibniz's side. Does not theodicy argue that an omnipotent God would allow evil to exist, because sin is unavoidable? That sin is not the agency of God but arises out of the necessary limitation of Man?

BONET: *(Shocked)* Argument? It's idle speculation of the worst kind. True heresy!

LADY BRASENOSE: Speculating about the existence of evil in a world created by a good God does not seem idle to me. Theodicy would claim that as Man cannot be absolutely perfect, Man's knowledge and power is limited. Thus we are not only liable to wrong action, but it is unavoidable or we would have absolutely perfect action from a less than absolutely perfect creature. How otherwise explain that God allowed Newton's manipulations? *(Pause)* Or do you attribute absolute perfection to Sir Isaac?

BONET: Lady Brasenose... now you are toying with me.

LADY BRASENOSE: Just returning to the issue at hand. Did you truly support Newton in an argument about the mathematical calculus by invoking religious reckoning... whatever that may be? There must be more.

BONET: *(Losing his temper)* There *is* more. The Royal Society honored me... a foreigner... by election to its illustrious fellowship. But my own

King's academy has not! Which academy's president would you support?

LADY BRASENOSE: Supporting Newton is unlikely to garner you election in Leibniz's academy.

BONET: But voting against Leibniz will.

LADY BRASENOSE: You have now piqued my curiosity.

BONET: You will hold this in confidence?

LADY BRASENOSE: If it merits such treatment.

BONET: Election to the theology class of our Academy is my desire. Its director, Daniel Ernst Jablonski, who also preaches at the King's Court, supports me. He founded the Academy with Leibniz. Whereas Leibniz as President receives a salary for life from the King, Jablonski and his colleagues receive nothing.

LADY BRASENOSE: Whereupon envy raised its ugly head!

BONET: Perhaps… but Leibniz's attention has since wandered far from the Academy.

LADY BRASENOSE: Meaning that a stipend for life is not further justified?

BONET: Putting it at the disposal of Jablonski… who labors night and day for the King's Academy… seems only just.

LADY BRASENOSE: And by reporting Newton's victory to your King, Leibniz's merits will diminish?

BONET: *(Admiringly)* My lady's acumen has not been blunted in six years.

LADY BRASENOSE: On the contrary, it has sharpened. *(Pause)* So as Minister, it will be your duty to report the Royal Society's conclusions to your court?

BONET: I shall dispatch a copy of the *Commercium Epistolicum* to Berlin. No commentary on my side will be necessary. It is damning enough.

LADY BRASENOSE: Because Newton wrote it.

BONET: No outsider is privy to that information.

LADY BRASENOSE: And your participation in the Committee?

BONET: Why disclose it when the Royal Society itself will not?

LADY BRASENOSE: *(Ironic)* My dear Bonet. You have just provided unimpeachable evidence in favor of theodicy…

END OF SCENE 9

Act 2, Scene 10

London 1731. Two comfortable club chairs, one occupied by DR JOHN ARBUTHNOT (now age sixty-four, in poor health, suffering from gout and kidney stones from which he will die within four years), his gouty leg on a low footstool and a cane by his side. COLLEY CIBBER (now sixty, but vigorous and active, and to live for another twenty-six years) is pacing.

ARBUTHNOT: Why? *(Pause, then with increasing intensity, though quivering voice)* Why? Why? Why? *(Longer pause)* Mr. Cibber! Why?

(CIBBER walks to the window, remains quiet, whereupon ARBUTHNOT resumes aggressive tone.)

Mr. Cibber, did you hear me?

CIBBER: Yes.

ARBUTHNOT: Then why not favor me with a reply?

CIBBER: These lines are in a play I know all too well. I wondered how much you intended to quote.

ARBUTHNOT: I was not quoting.

CIBBER: *(Slightly ironic tone)* Plagiarizing then?

ARBUTHNOT: Are you taunting me?

CIBBER: Only procrastinating.

ARBUTHNOT: Why? Guilt, perhaps?

CIBBER: I'm not accustomed to be questioned in so peremptory a manner.

ARBUTHNOT: Why then did you accept my invitation?

CIBBER: It was a summons... an imposition I generally ignore. What brought me here was sheer curiosity.... You wrote that you wanted to discuss a play... and that the matter was urgent.

ARBUTHNOT: It is urgent.

CIBBER: You did not mention the play's title.

ARBUTHNOT: And now you know.

CIBBER: So you saw it?

ARBUTHNOT: Last night.

CIBBER: The sixth performance... and still a full house.

ARBUTHNOT: A mob flocking into the theatre sheds little light on a play's quality... or veracity.

CIBBER: Since when is veracity on stage judged a virtue?

ARBUTHNOT: When it is not used to hide distortion.

CIBBER: Ours was applauded... your *Three Hours after Marriage* was hissed. Yours was virtually stillborn in 1717 and did not make it past the second performance. I know of no revival.

ARBUTHNOT: That is hitting below the belt.

CIBBER: Whose belt? John Gay's, Alexander Pope's, or yours? *(Scornfully)* Requiring three cooks for a thin theatrical pudding... meant to contain wit but in the end not tasting of wit at all. *(Short sarcastic laugh)* Asking the actors to do a good job while burdened with a bad script... meaning they had to be good at being bad!

ARBUTHNOT: Much too clever... and thus not worth recapture. You're more likely to be remembered for your sharper pen than for your tongue.

CIBBER: *(Prickly)* How so?

ARBUTHNOT: You had the audacity... some even called it impertinence... to adapt *Richard III*, but you added a line... "Off with his head... so much for Buckingham"... that I wager will be remembered longer than all the words you ever spoke on stage.

CIBBER: Is this a compliment or an affront?

ARBUTHNOT: The choice is yours! It is *Calculus* I wish to address... a true affront. Is a stage the place to wash dirty linen in public?

CIBBER: So we're back at veracity! Where else do such laundry? The stage is the only place where nothing need be hidden.

ARBUTHNOT: A country requires heroes—

CIBBER: *(Sarcastic)* Pray offer an example.

ARBUTHNOT: Take Marlborough.

CIBBER: A *military* hero, that I grant you. And awarded Blenheim Palace by a grateful sovereign and nation. But does that mean that John Vanbrugh—the architect of Blenheim—must also build a palace—a dramatic one—for Newton?

ARBUTHNOT: We need unsullied heroes... and not just military ones. What purpose is served by showing that England's greatest natural philosopher is flawed... like other mortals? Consider the laws of motion and of gravitation... of light and color... his work on celestial mechanics. Calculus was not needed for any of them. Even without the calculus, Newton would be our greatest.

CIBBER: Greatest natural philosopher... or paragon of probity? Why not take him for what he was: a tainted hero. Inventor of *the* calculus? Yes! But also corruptor of a moral calculus. And what about Leibniz... does he not deserve some defense?

ARBUTHNOT: Let that be the concern of the Germans.

CIBBER: Our Newton rests in Westminster Abbey under a hero's monument. But whatever their tomb, both continue to rot.

ARBUTHNOT: A medical or another moral judgment?

CIBBER: *(Conciliatory)* As you are a doctor, let it be medical. We've wrangled enough. *(Pause)* Poor Leibniz. The only person on the

Committee who could have defended him was a Swiss... and from Geneva at that!

ARBUTHNOT: Bonet. *(Reflects)* He had the least to lose... and thus lost the most.

CIBBER: You think Vanbrugh should have dealt with him more kindly?

ARBUTHNOT: Sir John is dead.

CIBBER: Sir John created the setting, he chose the characters, he dug up the dirt, and he spread it around. I only helped with broom and shovel... except for the very end. At his deathbed, Sir John asked me to complete the play... even offering me the epigraph: *frango ut patefaciam.*

ARBUTHNOT: "I break in order to reveal."

CIBBER: *(Nodding)* Your Latin is faultless. I acceded... with some reservation. The play was meant as revenge... though revenge, like love, is rarely consummated by surrogates. Yet directing retribution at the arbiters of our mores does suit me. Was I not also the object of their derision? Of course the theme Sir John chose—scandal among scholars and savants—is far removed from my experience. *(Pause)* But kindness is not a virtue in a play... nor are playwrights kind.

ARBUTHNOT: An expert speaking! But what about fairness? This is England... we have laws about fairness. *(Pause)* Consider libel.

CIBBER: I did. When Sir John died, Newton was eighty-four and ailing. I thought I'd wait—

ARBUTHNOT: For Newton to die?

CIBBER: The dead cannot be libeled... even if illuminating human frailty were considered ground for libel.

ARBUTHNOT: A legal opinion?

CIBBER: A logical one... in a country where the best laws often protect its worst people... and power and fame often do likewise.

ARBUTHNOT: So we are back to Newton.

CIBBER: We could not have had *Calculus* without Newton.

ARBUTHNOT: Of course. He discovered it.

CIBBER: But so did Leibniz... and published it first. Therefore, the calculus would be with us... even without Newton. But our play? Vanbrugh was right: deepest corruption... and thus vilest scandal... is intellectual—not sexual. The play is about Newton's malfeasance.

ARBUTHNOT: Yet the mirror you use is the Committee.

CIBBER: Well put, Dr. Arbuthnot!

ARBUTHNOT: You did not just complete the play, you played in it!

CIBBER: I'm an actor as well as writer.

ARBUTHNOT: A better actor than an author.

CIBBER: *(Aside)* A judgment I've heard before.

ARBUTHNOT: Since I was on the Committee—

CIBBER: You were also in our play.

ARBUTHNOT: Hardly as a minor character!

CIBBER: Is this a cause for complaint?

ARBUTHNOT: A major one... considering how you depict me. *(Angrily)* I'm still alive!

CIBBER: And brimful of vigor as you just demonstrated.

ARBUTHNOT: But in terrible health! I suffer deeply from mysterious fevers and a great stone in my right kidney... a punishment for years of overindulging my palate. And now the gout! *(Grimacing, points to his foot with his cane)*

CIBBER: *(Trying to be conciliatory)* I'm sure you judge yourself too severely.

ARBUTHNOT: *(Ignores the comment)* I have buried six of my children and recently my wife... and now find my reputation buried as well!

CIBBER: *(Uncomfortable)* Please accept my condolences—

ARBUTHNOT: From you... who lashed me with a whip?

CIBBER: A moral whip… and only in a play.

ARBUTHNOT: And therefore worse… with exposure all too public and thus with pain that much greater. But was it justified? Where did you learn the facts you purport to describe?

CIBBER: From Sir John.

ARBUTHNOT: And he?

CIBBER: I suspect from Lady Brasenose.

ARBUTHNOT: And she?

CIBBER: From various sources… for instance Bonet.

ARBUTHNOT: The play is about Newton… sage and genius—

CIBBER: *(Interrupts quickly)* And his faults—real or perceived—as well as sins or even crimes!

ARBUTHNOT: But how do you know that Lady Brasenose had met Bonet?

CIBBER: Because…

ARBUTHNOT: Yes?

CIBBER: Because… *(Pause)*… because she said so.

ARBUTHNOT: You heard her say so?

CIBBER: Our paths have never crossed. She told Sir John.

ARBUTHNOT: He said so?

CIBBER: I assumed… because he so implied.

ARBUTHNOT: Your assumption about the implication is wrong. All Lady Brasenose knew about Bonet she learned from someone else.

CIBBER: *(Defensive)* And Moivre?

ARBUTHNOT: *(Sarcastic)* What did she learn from him… in *Calculus*? That he was poor? Every Fellow of the Royal Society knew of his poverty... and those that could have helped him overcome it... didn't... not to this day.

CIBBER: *(Even more defensive)* He told her about fluxions... and calculus... and—

ARBUTHNOT: *(Short sardonic laugh)* Mathematics? There is precious little about it in your *Calculus*. But why should there be? It is about *mathematicians*... not mathematics. About Newton and Leibniz... or Halley and Keill or—

CIBBER: You?

ARBUTHNOT: I am not of their scholarly rank... nor did I favor any of their conflicts. I prefer to resolve discord... sometimes even at great personal cost. Yet none of them appears on stage... whereas I do! Are actors not supposed to *show* rather than *tell*? *(Pause)* Newton or Leibniz and the rest show nothing. What we learn all leads to Lady Brasenose.

CIBBER: True... even the infamous anagram.

ARBUTHNOT: *(Dismissive)* Oh yes... anagrams! *(Affected tone)* As in "*Calculus: a Morality Play* by H. Van Grub and Colley Cibber." *(Dismissive)* How utterly transparent!

CIBBER: Sir John had planned to use an alias, but left none upon his death. I chose "H. Van Grub"... and thought it clever.

ARBUTHNOT: "Grub" as in drudge or hack? Or Grub Street where the hacks and drudges live? Appropriate? Perhaps. But clever?

CIBBER: You think of the noun... I of the verb. To "grub" is to dig... and usually dig for dirt. That is what *Calculus* is about.

ARBUTHNOT: In that case... back to the dirt. She did not know Bonet or Moivre. Who provided Lady Brasenose with the clues?

CIBBER: Surely not you?

ARBUTHNOT: My wife... and therefore I.

CIBBER: *(Dumbfounded)* But why?

ARBUTHNOT: I've often ridiculed pretentious erudition and scholarly jargon with Pope, with Swift, with Parnell, and with Gay... even forming a club, the Scriblerus. But Newton's and Leibniz's erudition was neither pretense nor their scholarly dispute jargon. Much of it was poison that demeaned both. Ridicule was not a cure. I tried compromise and reason... yet in the end failed. And since all parties die at last of swallowing their own lies...

CIBBER: I seem to have read that somewhere.

ARBUTHNOT: It's from *The Art of Political Lying*... a book I wrote myself.

CIBBER: Self-quotation does not guarantee veracity.

ARBUTHNOT: Nor exclude it. I felt a serious message was indicated—a form of moral revenge.

CIBBER: You sound like Vanbrugh.

ARBUTHNOT: Sir John is dead, but were he alive, I suspect he would concede that I came first. Why not a play... a morality play... but suitably disguised and libel-proof... to teach a lesson?

CIBBER: You thought of that?

ARBUTHNOT: And told a well-connected person... curious and moral... about my plan.

CIBBER: Lady Brasenose?

ARBUTHNOT: I even suggested the title. After all, everyone was calculating in one way or another... even the ones who knew no calculus.

CIBBER: When I asked Sir John why he insisted on so tame a title—

ARBUTHNOT: So you thought it tame?

CIBBER: *(Retracts)* Call it too mysterious. On this point, however, Sir John was immovable.

ARBUTHNOT: I take it you had another preference.

CIBBER: It was "Newton's Whores." A title only slightly longer than "Calculus" but much more beguiling.

ARBUTHNOT: I'd consider it too ambiguous. Which horse is Newton riding? *(Perhaps even mimes clumsily riding a horse)* Galloping with the calculus... or trotting with the committee's conclusions?

CIBBER: I was referring to humans ... not equines.

ARBUTHNOT: *(Spells it out slowly)* W-H-O-R-E-S?

CIBBER: *(Mimes deferential bow)* How perceptive!

ARBUTHNOT: *(Primly)* For all too obvious reasons, I find such a title excessively offensive... even for your play.

CIBBER: *(Defensive)* There are varieties of whoring applicable to men: traders of flattery... begetters of lies... spreaders of gossip—the toadies of this world... the Rosencrantz and Guildenstern characters... whom we actors know all too well... and find again in this play. When Sir John spoke of revenge... he meant revenge by shedding light upon such men and a society that fosters them. By showing how even small incremental changes over time... call them fluxions in our behavior... lead to measurable conclusions. *(Pause)* But truly... not just the title, but the play as well was your idea?

ARBUTHNOT: *(Nods)* Of course I could not write the play. *(Sarcastic)* As Newton preached, though never practiced, "no man is a witness to his own cause." But was not Vanbrugh skilled in writing plays about real persons well disguised? Why not ask him? I asked Lady Brasenose. "I shall do it," she exclaimed and then did so... feeding him judiciously... step-by-step... plotless information... and even a title. The rest... you know.

CIBBER: I am dumbfounded.

ARBUTHNOT: And so was I... last night at Drury Lane. I expected a play that even Newton could have seen. Of course, not liked... but seen... because the author's subtlety would have prevented open accusations. Yet one of the authors shielded himself through an alias—a privilege he did not extend to his victims. Morality plays should teach a lesson the accused can witness. I wanted to wound Newton without leaving a mark. But with your *Calculus...* to besmirch him permanently... you had to wait for Newton's burial.

CIBBER: And Vanbrugh's.

ARBUTHNOT: Surely he did not plan that?

CIBBER: He knew he was close to the end... but not that close. He never told me how he'd end the play.

ARBUTHNOT: You could have changed it... and much else. Why didn't you?

CIBBER: There is a line in my first play—

ARBUTHNOT: *Love's Last Shift*?

CIBBER: The very one, where I wrote: "The world to me is a garden stocked with all sorts of fruit, where the greatest pleasure we can take is in the variety of taste." I've never tasted such a fruit in the theatre... so why alter it?

ARBUTHNOT: I wanted somebody to write a play about the *cost* of destroying reputations... whereas *(Sarcastic)*... "H. Van Grub" and you simply chose to destroy reputations... whatever the cost incurred.

CIBBER: Perhaps we all miscalculated.

ARBUTHNOT: Perhaps we did.

(Painfully rises, leaning heavily on his cane and starts hobbling away.)

THE END

$\mathcal{A}$uthors' Biographical Sketches

Carl Djerassi

Carl Djerassi, novelist, playwright and Professor of Chemistry Emeritus at Stanford University, is one of the few American scientists to have been awarded both the National Medal of Science (for the first synthesis of an oral contraceptive) and the National Medal of Technology (for promoting new approaches to insect control). Among numerous other recognitions, he is a member of the National Academy of Sciences as well as the American Academy of Arts and Sciences and the recipient of 19 honorary doctorates. Aside from nine monographs and over 1200 scientific articles, he has published short stories *(The Futurist and Other Stories)*, poetry *(The Clock Runs Backward)* and five novels *(Cantor's Dilemma; The Bourbaki Gambit; Marx, deceased; Menachem's Seed; NO)*—that illustrate as "science-in-fiction" the human side of science and the personal conflicts faced by scientists—as well as an autobiography *(The Pill, Pygmy Chimps and Degas' Horse)* and a memoir *(This Man's Pill: Reflections on the 50th Birthday of the Pill)*.

During the past six years he has focused on writing "science-in-theatre" plays. The first, *An Immaculate Misconception*, premiered at the 1998 Edinburgh Fringe Festival and was subsequently staged in London (New End Theatre in 1999 and Bridewell Theatre in 2002), San Francisco (Eureka), New York (Primary Stages), Vienna (Jugendstiltheater), Cologne (Theater am Tanzbrunnen), Munich (Deutsches Museum), Sundsvall (Teater Västernorrland), Stockholm (Dramaten), Sofia (Satire Theatre), Geneva (Theatre du Grütli), Tokyo (Bunkyo Civic Hall Theatre) and Seoul (Dong Duk Womans University Art Center). The play has been translated into eight languages and also been published in book form in English, German, Spanish, and Swedish.

The BBC broadcast the play in 2000 as "play of the week" on the World Service, and the West German Rundfunk (WDR) and Swedish Radio did so in 2001.

His second play, *Oxygen*, co-authored with Roald Hoffmann, premiered in April 2001 at the San Diego Repertory Theatre, at the Mainfranken Theater in Würzburg in September 2001 through April 2002 (with guest performances in 2001/2002 in Munich, Leverkusen and Halle), at the Riverside Studios in London in November 2001, and in Korea in August 2002. New Zealand (Circa Theatre, Wellington), Japanese (Setagaya Tram Theatre, Tokyo), Italian (Bologna) and Bulgarian (Sofia, Satire Theatre) premieres are scheduled for 2003. Both the BBC and the WDR broadcast the play in December 2001 around the centenary of the Nobel Prize—one of that play's main themes. It has so far been translated into 7 languages with 3 others underway and has already appeared in book form in English, German, Italian and Korean.

His third play, *Calculus*, dealing with the infamous Newton-Leibniz priority struggle, has had staged rehearsed readings in Berkeley, London (Royal Institution), Vienna, Munich and Oxford (Oxford Playhouse) and opened in San Francisco (Performing Arts Library and Museum) in April 2003. His first "non-scientific" play, *Ego,* premiered at the 2003 Edinburgh Festival Fringe.

Djerassi is the founder of the Djerassi Resident Artists Program near Woodside, California, which provides residencies and studio space for artists in the visual arts, literature, choreography and performing arts, and music. Over 1200 artists have passed through that program since its inception in 1982. Djerassi and his wife, the biographer Diane Middlebrook, live in San Francisco and London.

(There is a Web site about Carl Djerassi's writing at http://www.djerassi.com.)

David Pinner

David Pinner was born in Peterborough, England, and was trained as an actor at the Royal Academy of Dramatic Art. For the last ten years he has been Visiting Associate Professor of Drama at Colgate University in New York. As an actor he has played leading roles at the Mermaid, Stratford East, the North East Festival, the Theatre Royal, Sheffield and the Theatre Royal, Windsor. In television he has appeared in many roles on ITV and the BBC, including William in *Henry V* and the Duke of Clarence in *The Prince Regent*. While playing the lead in *The Mousetrap* in the Ambassador's Theatre, West End, he wrote his first novel *Ritual* upon which the cult movie *The Wicker Man* was based. He has written two other novels: *With My Body* and *There'll Always Be An England.*

Most of his writing has focused on plays for the theatre, which include *Fanghorn* with Glenda Jackson (Fortune Theatre, London); *Lucifer's Fair* (Arts Theatre), and his *Stalin Trilogy:* the *Potsdam Quartet* (Lyric Theatre, Hammersmith and Lion Theatre, New York), which was screened by the BBC; *The Teddy Bears' Picnic* (Gateway Theatre, Chester) and *Lenin in Love* (New End Theatre, Hampstead). Other plays include *The Last Englishman* (Orange Tree, Richmond); *Shakebag, An Evening With the G.L.C.* and *Cartoon* (Soho Poly), and *Sins of the Mother* (Grace Theatre). He has had eighteen plays produced, ten plays published, and had many of them broadcast on BBC radio, and BBC and commercial television. His television film *Juliet and Romeo* was shown on German television. His *Revolution Trilogy* comprising *The Drums of Snow*, *Cardinal Richelieu* and *Talleyrand, Prince of Traitors* will be published in 2003.

Pinner has directed many plays in the UK and the USA, most recently (2002) his *All Hallow's Eve* at the Hong Kong Academy of Performing Arts and his *Midsummer* in the USA. He has just completed a musical on Marx and Engels; *Marx and Sparks.*

(There is a Web site about David Pinner's writing at http://groups.colgate.edu/klatsch/archiv/pinner_folder/dphpg3.html.)

*A*cknowledgments

Carl Djerassi's acknowledgments—typical of an ex-academic—are long and start with money: I am indebted to the office of the Dean for Undergraduate Education at Stanford University for funds to enable two of my top students to assist in long-term research on historical background for this play. Joshua Bushinsky provided crucial evidence on the eleven fellows of the Royal Society that made up the infamous committee appointed to adjudicate the Newton-Leibniz controversy, while Tonyanna Borkovi conducted extensive research on important background relating to Colley Cibber, John Vanbrugh and other theatre personalities of the Restoration period.

Professor Walter Grünzweig of the University of Dortmund, Professor Armand Buchs of the University of Geneva, Professor Peter Laur of the Technical University, Aachen, Dr. Gudrun Pisching of the Karl Franzen University, Graz, and Dr. Andrew Thompson of Queens' College, Cambridge offered crucial leads to the least known character in *Calculus*—Louis Frederick Bonet, the King of Prussia's Minister to England—which enabled me to construct a plausible motive for his puzzling role as foreign member of the Royal Society's committee.

Alan Drury, formerly dramaturg of BBC Radio's theatre department, and Prof. Diane Middlebrook of Stanford University, wielded sharp critical stilettos that persuaded me to kill many a literary darling and even some marvelous, though esoteric, tidbits that authors of historical events are usually loath to destroy. And finally, it is my sad duty to acknowledge the enormously sympathetic comments of Robert Merton, Professor Emeritus of Sociology at Columbia University—the unchallenged dean of sociologists of science and author of a book *(On the Shoulders of Giants)* that no Newton scholar can ignore—who died a few months before the American premiere of *Calculus*. Nothing would have pleased me more than to have had his company at the first staging of *Calculus*.

David Pinner's acknowledgment is pithy—as behooves a professional playwright: I have written *Newton's Hooke* in close collaboration with my son, Dickon, who is a physicist. It was Dickon's idea that I should write a play about Isaac Newton and Robert Hooke. His inspiration, wisdom and guidance have played a pre-eminent part in the creation of this play.

Finally, both authors wish to express their indebtedness to Shannon Moffat of Stanford University and Iolanda Antunes of San Jose for crucial assistance in the generation of the final, camera-ready manuscript.